DISCLAIMER

This book is intended for i.............. purposes only. It does not constitute medical advice, diagnosis, or treatment, and should not be relied upon as a substitute for professional medical advice from a qualified healthcare provider. Always consult your physician or other qualified health professional before starting any new diet, health plan, or herbal regimen, especially if you have any pre-existing medical conditions, are taking medications, are pregnant, or are breastfeeding. The author and publisher disclaim any liability for adverse effects resulting from the use or application of the information contained in this book. By reading this book, you acknowledge that you are responsible for your own health decisions and agree to hold the author and publisher harmless.

CONTENTS

CHAPTER 1 10

THE WORLD OF RAGDOLL CATS 10

1.1 Introduction to Ragdolls *10*

1.2 Origins and Development of the Breed.............*11*

1.3 Myths and Misconceptions.............................*14*

1.4 Why Ragdolls Are Called "Gentle Giants".......*16*

CHAPTER 2 19

HISTORY AND BREED OVERVIEW 19

2.1 The Creation of the Ragdoll Breed..................*19*

2.2 Recognition by Cat Associations *21*

2.3 Breed Standards and Physical Traits...............*22*

2.4 Popularity and Cultural Impact......................*23*

CHAPTER 3 25

PERSONALITY AND SUITABILITY AS PETS.....................25

3.1 Temperament and Personality Traits *25*

3.2 Ragdolls vs. Other Cat Breeds........................*27*

3.3 Are Ragdolls the Right Choice for You? *29*

3.4 Common Misunderstandings About Their Behavior *31*

CHAPTER 4 35

PREPARING YOUR HOME FOR A RAGDOLL.....................35

4.1 Safe Indoor Environment Setup *35*

4.2 Essential Living Spaces for Cats *37*

4.3 Cat-Proofing Your Home.................................*39*

4.4 Creating a Stress-Free Transition...................*40*

CHAPTER 5 43

RAGDOLL CAT

Master The Art of Ragdoll Cat Handbook – Feeding, Grooming, Training, Behavior, and Health Tips for Gentle Giants

MARK J. PETERSON

FEEDING BASICS 43

5.1 Understanding Nutritional Needs................................43

5.2 Choosing Balanced Meals....................................45

5.3 Feeding Schedules and Portions 47

5.4 Avoiding Common Feeding Mistakes.........................49

CHAPTER 6 ....**52**

HOUSING AND SPACE NEEDS...52

6.1 Indoor Living for Ragdolls 52

6.2 Ideal Sleeping Areas and Rest Zones 54

6.3 Multi-Pet Households Considerations........................55

6.4 Outdoor Access and Safety Concerns 57

CHAPTER 7 ....**60**

GROOMING AND COAT CARE 60

7.1 Brushing and Detangling Ragdoll Fur.......................60

7.2 Bathing: When and How....................................62

7.3 Nail Care and Paw Maintenance 63

7.4 Eye, Ear, and Dental Hygiene 65

CHAPTER 8 ....**68**

TRAINING FUNDAMENTALS 68

8.1 Why Training Matters for Cats 68

8.2 Positive Reinforcement Techniques 69

8.3 Teaching Boundaries and Rules............................71

8.4 Litter Training and Success Tips............................73

CHAPTER 9 ....**76**

SOCIALIZATION AND BONDING 76

9.1 Helping Your Cat Adapt to New People 76

9.2 Bonding Activities for Trust Building........................78

9.3 Introducing Ragdolls to Other Pets......................................79

9.4 Creating a Loving Daily Routine.....................................81

CHAPTER 10 ....**84**

UNDERSTANDING RAGDOLL BEHAVIOR 84
10.1 Natural Instincts of Ragdolls....................................84

10.2 Recognizing Body Language 86

10.3 Affectionate vs. Independent Behaviors 88

10.4 Signs of Stress or Discomfort 89

CHAPTER 11 ....**92**

PLAY, ENRICHMENT, AND STIMULATION.............................92
11.1 Importance of Play for Ragdolls 92

11.2 Types of Enrichment Activities................................93

11.3 Interactive Play for Owner-Cat Bond 95

11.4 Preventing Boredom and Stress...............................96

CHAPTER 12 ....**99**

HEALTH BASICS 99
12.1 Routine Veterinary Visits 99

12.2 Vaccinations and Preventive Care 101

12.3 Common Health Issues in Ragdolls........................102

12.4 Spotting Early Warning Signs 104

CHAPTER 14 ....**107**

GROWTH AND LIFE STAGES..107
14.1 Kitten Development Stages 107

14.2 Caring for Adolescent Ragdolls...............................109

14.3 Adult Cat Care Essentials.......................................110

14.4 Senior Cat Adjustments 112

CHAPTER 15 ....**115**

HANDLING AND INTERACTION 115

15.1 Correct Ways to Pick Up and Hold Ragdolls 115

15.2 Encouraging Gentle Handling for Children 117

15.3 Reducing Fear During Vet Visits.................................118

15.4 Stress-Free Travel Tips 120

CHAPTER 16 ....**123**

SAFETY AND HOUSEHOLD HAZARDS...123

16.1 Common Indoor Dangers for Cats 123

16.2 Toxic Plants and Foods 125

16.3 Preventing Accidents in the Home.............................127

16.4 Fire, Emergency, and Evacuation Preparedness........129

CHAPTER 17 ....**132**

BREEDING BASICS (OPTIONAL KNOWLEDGE) 132

17.1 Understanding Ragdoll Genetics 132

17.2 Responsible Breeding Ethics 134

17.4 Raising Healthy Litters...137

CHAPTER 18 ....**139**

VETERINARY CARE AND COMMON ILLNESSES.....................................139

18.1 Genetic Predispositions in Ragdolls 139

18.2 Common Illnesses and Treatments............................141

18.3 Preventing Obesity and Joint Issues.........................143

18.4 Supporting Long-Term Health 145

CHAPTER 20 ....**147**

LIFESPAN AND END-OF-LIFE CARE 147

20.1 Average Lifespan of Ragdolls....................................147

20.2 Comforting Senior Cats...150

20.3 Palliative and End-of-Life Support 153

20.4 Coping with Loss and Grief 157

CHAPTER 21 ....**162**

BEHAVIORAL CHALLENGES 162

21.1 Separation Anxiety Solutions 162

21.2 Aggression and Biting...164

21.3 Over-Grooming and Stress Behaviors.......................165

21.4 Litter Box Avoidance Fixes 167

CHAPTER 22 ....**170**

ADVANCED TRAINING AND ENRICHMENT................................170

22.1 Teaching Tricks and Commands 170

22.2 Harness and Leash Training....................................171

22.3 Mental Enrichment Exercises..................................173

22.4 Problem-Solving for Stubborn Cats..........................174

CHAPTER 23 ....**177**

250+ BONUS TIPS, HACKS, AND FAST FACTS 177

23.1 Fun Ragdoll Trivia 177

23.2 Quick Health and Grooming Hacks...........................178

23.3 Everyday Time-Saving Tricks..................................179

23.4 Enrichment and Bonding Boosters 180

CHAPTER 24 ....**182**

FREE 30-DAY CARE & TRAINING PLAN 182

24.1 Week 1: Settling in and Bonding.............................182

24.2 Week 2: Feeding, Grooming, and Play.......................184

24.3 Week 3: Training and Socialization 186

24.4 Week 4: Building Long-Term Routines.......................188

CHAPTER 25 191

CONCLUSION & KEY TAKEAWAYS 191

25.1 Lessons Learned from Ragdoll Ownership 191

25.3 Final Care Reminders and Essentials 194

25.4 Words of Encouragement for New Owners 195

CHAPTER 1

The World of Ragdoll Cats

1.1 Introduction to Ragdolls

Ragdoll cats are one of the most charming, affectionate, and strikingly beautiful feline breeds in the world. Often described as "dog-like" in their loyalty and sociability, Ragdolls are renowned for their tendency to follow their humans from room to room, their soft and silky coats, and their striking blue eyes that seem to glow with personality. For first-time cat owners or those looking for a gentle, easy-going feline companion, the Ragdoll presents an ideal choice.

Ragdolls are not your average house cat. Unlike many feline breeds that pride themselves on independence and aloofness, Ragdolls are interactive, human-centered, and affectionate to a fault. They enjoy being cuddled, carried, and involved in the daily activities of their owners. In fact, many Ragdolls owners report that their cats greet them at the door, sleep beside them at night, and engage in affectionate gestures like head-butting, slow blinks, or gentle purring while making eye contact.

One of the first things people notice about Ragdolls is their size. They are among the largest domesticated cat breeds, with males weighing anywhere from 15 to 20 pounds, and females typically between 10 to 15 pounds. But it's not just their size that earns them admiration — it's their luxurious coat, placid expression, and gentle nature that make them beloved among cat fanciers and pet owners alike.

Their name, "Ragdoll," stems from their tendency to go limp when picked up. Although this trait is somewhat exaggerated and not true of every individual cat, many Ragdolls do relax significantly when held, giving the impression of a plush toy. This relaxed behavior is part of their appeal and reflects their trusting and non-aggressive temperament.

Key Takeaway: Ragdolls are calm, loving, and social cats that thrive in family homes. Their interactive personalities and gentle behavior make them a perfect choice for beginners and experienced pet owners alike.

1.2 Origins and Development of the Breed

The story of the Ragdoll breed begins in the 1960s in Riverside, California, where a woman named Ann Baker began selectively breeding cats to develop a new and distinctive breed. At the center of the Ragdoll's origin is a white domestic longhaired cat named Josephine. According to Baker, Josephine had an unusually gentle temperament, and following a car accident, she gave birth to kittens with similarly docile and affectionate natures.

Ann Baker believed that the kittens had inherited not only Josephine's physical beauty but also her remarkable behavior. Whether or not the car accident truly changed Josephine's genetics — a claim viewed with skepticism by scientists — Baker began carefully breeding Josephine's offspring with other cats who exhibited desirable traits. She was particularly interested in preserving the relaxed, easy-going personality and luxurious coats of these cats.

Using cats such as Burmese, Birman, and Persian types, Baker created a new breed that combined striking beauty with a placid nature. She named this breed "Ragdoll," both as a nod to their relaxed demeanor and to distinguish the new line from any existing breed.

In an unusual move, Ann Baker trademarked the name "Ragdoll" and established the International Ragdoll Cat Association (IRCA), where she set strict breeding guidelines and retained legal control over the breed. She limited breeders' ability to register Ragdolls with other cat associations, causing tension in the breeder community. Her approach, while ensuring a consistent standard early on, ultimately caused division.

It wasn't long before a group of breeders broke away from IRCA, opting instead to pursue formal recognition of the breed with major cat registries. This split led to the broader availability and acceptance of Ragdolls as a legitimate breed outside of Baker's tight control.

Despite early controversies, Ann Baker's vision and pioneering efforts laid the groundwork for what would become one of the most beloved cat breeds worldwide.

Key Takeaway: Ragdolls were created through selective breeding for affectionate, relaxed personalities. Their development was rooted in one woman's vision, which laid the foundation for a breed that would go on to achieve international acclaim.

1.3 Myths and Misconceptions

As with many popular breeds, Ragdolls have been the subject of various myths and misconceptions. While some of these beliefs are harmless exaggerations, others can lead to misunderstanding or even improper care if accepted as fact. This section helps debunk the most common myths and offers clarity for first-time Ragdoll owners.

Myth #1: Ragdolls don't feel pain

This dangerous myth likely stems from their relaxed nature and tendency to stay calm in situations where other cats might react more strongly. In truth, Ragdolls experience pain just like any other cat. Their stoic nature may cause them to exhibit fewer outward signs of discomfort, making it important for owners to monitor their health closely and not ignore subtle signs of illness or injury.

Myth #2: All Ragdolls go limp when picked up

The "ragdoll flop" is one of the breed's signature traits, but it's not universal. While many Ragdolls do relax in their owner's arms, not all will display this behavior. Each cat is unique, and temperament can vary depending on genetics, early handling, and individual personality.

Myth #3: Ragdolls are hypoallergenic

While Ragdolls produce less of the Fel d 1 protein than some other cats, they are not hypoallergenic. Individuals with severe cat allergies should spend time with a Ragdoll before bringing one home to see how their body reacts.

Myth #4: Ragdolls don't need grooming

Because Ragdolls lack an undercoat, their fur is less prone to matting. However, this does not mean they are maintenance-free. Regular grooming is essential to keep their coats healthy and reduce shedding. Owners should still brush their Ragdoll a few times per week.

Myth #5: Ragdolls are lazy

While Ragdolls are laid-back, they are not inactive. They enjoy playtime and require mental and physical stimulation like any other cat. Enrichment activities, interactive toys, and human interaction are important for their well-being.

Key Takeaway: Ragdolls are often misunderstood due to their unique temperament. Knowing the facts helps owners provide better care and avoid common pitfalls.

1.4 Why Ragdolls Are Called "Gentle Giants"

The nickname "gentle giant" captures the very essence of the Ragdoll breed. They are among the largest domestic cat breeds and are simultaneously among the most affectionate and docile. This seemingly paradoxical combination of size and sweetness is what makes Ragdolls so special.

Physical   **Traits**

Ragdolls are large cats with strong, muscular bodies and long legs. Their tails are bushy, and their coats are thick but silky. Males often weigh between 15 to 20 pounds, and females range from 10 to 15 pounds. Despite their size, they move gracefully and with purpose.

Their eyes are another defining feature: large, oval, and deep blue, conveying both beauty and a sense of deep awareness. Their ears are medium-sized, rounded at the tips, and slightly tilted forward, giving them an inquisitive, alert appearance.

Temperament and Personality

What truly sets Ragdolls apart, however, is their temperament. They are not aggressive, territorial, or independent in the way many cats are. Instead, they crave affection and physical closeness. They often sleep next to or on their humans, enjoy being carried like a baby, and love gentle petting and soothing talk.

This gentle nature makes them well-suited for families with children, older adults, and even multi-pet households. Ragdolls are typically non-aggressive with other animals, and their calm demeanor makes them easy to introduce into new environments.

Their Role as Emotional Companions

Because of their consistent and predictable behavior, Ragdolls are often viewed as therapeutic companions. Many owners describe their Ragdolls as being attuned to their emotional states, often providing comfort and quiet support simply by being present. Their calm presence has been noted as beneficial for individuals dealing with stress, anxiety, or loneliness.

Living with a Gentle Giant

Daily life with a Ragdoll means sharing space with a loyal, mellow friend who enjoys being wherever you are. They don't hide under furniture or shy away from company; they greet visitors, lie beside you on the couch, and even perch on your desk as you work.

Key Takeaway: Ragdolls are large cats with loving hearts. Their nickname "gentle giant" reflects not just their size but their affectionate, people-focused nature. They are calm, comforting companions that add warmth and tranquility to any home.

CHAPTER 2

History and Breed Overview

2.1 The Creation of the Ragdoll Breed

The Ragdoll cat breed traces its origins to a single breeder's vision and commitment to developing a feline that was as affectionate as it was beautiful. In the early 1960s, Ann Baker, a Persian cat breeder living in Riverside, California, noticed a unique trait in a white longhaired cat named Josephine. After Josephine was involved in an automobile accident and subsequently nursed back to health, her litters began displaying peculiar behavior: extreme docility, calmness, and an unusual tolerance for being handled. Ann believed the trauma may have caused genetic changes, though this theory has no scientific basis.

Nevertheless, Baker saw potential in Josephine's kittens. She began breeding Josephine with other cats who displayed similar behaviors and physical characteristics. This included a seal mitted male named Daddy Warbucks and a solid black cat named Blackie. Through a carefully orchestrated breeding program, Baker selectively refined the characteristics of the kittens to be large, blue-eyed, semi-longhaired, and incredibly gentle.

What set Baker apart was not just her breeding success, but also her strategy. She trademarked the name "Ragdoll" and founded her own registry: the International Ragdoll Cat Association (IRCA). This move allowed her to maintain exclusive rights and strict control over who could breed Ragdolls and how.

While the legitimacy of Baker's genetic claims remains debated, her contributions to establishing the foundation of the breed are undeniable. Her focus on behavior as much as physical traits helped shape the Ragdoll into a unique type of companion animal.

Key Takeaway: The Ragdoll breed was the result of selective breeding for calm temperament and beauty, pioneered by Ann Baker's innovative and controversial methods.

2.2 Recognition by Cat Associations

Following the early development of the breed under Ann Baker's IRCA, a number of breeders grew disillusioned with the restrictive nature of her control. Seeking a more open path, Denny and Laura Dayton were among the first to branch away from IRCA, working to establish breed standards that would allow Ragdolls to be recognized by traditional cat associations.

Their efforts were successful. By the 1970s and 1980s, the Ragdoll breed began to gain acceptance from major cat registries. TICA (The International Cat Association) recognized the breed in the 1980s, and CFA (Cat Fanciers' Association) followed in the early 1990s. Each organization had its own criteria for recognition, but all valued the same essential features: a striking, large, blue-eyed, pointed breed with a calm and placid personality.

Recognition meant more than prestige. It meant breeders could now show their Ragdolls at official cat shows, access a wider breeding network, and ensure that Ragdolls were bred to consistent, documented standards. Today, Ragdolls are fully recognized by nearly every major feline organization around the world.

Key Takeaway: Recognition by major cat associations helped transform Ragdolls from a niche experiment to a globally respected breed.

2.3 Breed Standards and Physical Traits

The Ragdoll breed is distinguished by a precise set of physical characteristics outlined by each major cat registry. While there are small differences among organizations, the core traits remain consistent:

- **Size:** One of the largest cat breeds. Males: 15–20 lbs; Females: 10–15 lbs.

- **Coat:** Silky, semi-longhair with no undercoat. Requires brushing 2–3 times per week.

- **Eyes:** Always blue. Shape is large and oval.

- **Ears:** Medium-sized, rounded at the tips, and slightly forward-tilted.

- **Color Patterns:** Typically, Colorpoint, Mitted, or Bicolor. Common colors include seal, blue, lilac, and cream.

In addition to their looks, Ragdolls are prized for their demeanor. Most standards require a friendly, easy-going, and affectionate personality. In shows, judges even assess how calm the cat remains when being handled — a nod to the Ragdoll's reputation as a "people cat."

Key Takeaway: Ragdolls are not only beautiful — they meet specific breed standards that emphasize both physical elegance and behavioral calm.

2.4 Popularity and Cultural Impact

Today, Ragdolls consistently rank among the top five most popular cat breeds in the United States, United Kingdom, and Europe. Their celebrity status has been cemented not just by their beauty and temperament, but by their adaptability to different households and lifestyles.

They are frequently featured in cat calendars, social media posts, and pet commercials. Ragdoll influencers have emerged on platforms like Instagram and TikTok, attracting millions of followers who love watching these cats lounge, play, and cuddle with their families. Their calm demeanor also makes them a favorite breed for therapy cats and feline-assisted emotional support roles.

Additionally, Ragdoll-specific breed clubs, shows, and rescue organizations have grown around the world. These communities promote breed education, advocate for ethical breeding practices, and provide resources for new and potential owners.

Key Takeaway: The Ragdoll's rise to fame is fueled by its gentle nature and widespread appeal, making it a fixture in modern pet culture and a favorite among fami

CHAPTER 3

Personality and Suitability as Pets

3.1 Temperament and Personality Traits

Ragdoll cats are widely celebrated for their gentle, affectionate, and calm personalities. Unlike many feline breeds that exhibit strong independence, Ragdolls are known for their deep attachment to their human companions. They enjoy being physically close—whether sitting beside you, lying on your lap, or following you around the house. This is often described as "puppy-like" behavior, a unique characteristic that sets them apart from more aloof cat breeds.

Ragdolls are typically quiet and soft-spoken. Their voices are subtle, rarely vocalizing loudly unless there's a specific need such as hunger or attention. They tend to communicate through body language—such as nudging with their head, gently pawing at you, or rubbing their body against yours. These physical gestures are strong indicators of their desire to connect and bond with their humans.

They're also extremely tolerant of handling. Ragdolls are one of the few breeds that genuinely enjoy being picked up and carried. They rarely scratch or lash out, even when slightly annoyed, making them well-suited for families with children or multiple pets. Their easy-going nature also extends to grooming and vet visits, where they often remain calm and cooperative.

Although laid-back, Ragdolls are still playful and enjoy interactive toys, games of fetch, and puzzle feeders. They are intelligent, capable of learning simple tricks, and enjoy being mentally stimulated through games and engagement. However, they prefer gentle play and may shy away from rough or fast-paced activities.

Another defining personality trait of Ragdolls is their adaptability. They adjust well to new homes and tend to stay composed during environmental changes, such as moving or the introduction of new family members. This adaptability contributes to their overall appeal as low-stress pets.

Additionally, Ragdolls are not known to be climbers. While many cats enjoy perching atop cabinets or shelves, Ragdolls prefer to stay grounded. They may climb a cat tree or hop on the couch, but they're not typically drawn to high spaces. This grounded nature makes them well-suited to apartment dwellers or households with limited vertical space.

Key Takeaway: Ragdolls are affectionate, easy-going, adaptable cats with gentle temperaments, making them ideal companions for people seeking a deeply bonded feline friend.

3.2 Ragdolls vs. Other Cat Breeds

While many cat breeds possess unique and admirable traits, Ragdolls stand out for their combination of large size, docile behavior, and social nature. This section compares Ragdolls to a few other popular breeds, highlighting the differences in personality, care needs, and lifestyle compatibility.

Ragdoll vs. Maine Coon: Both breeds are large and affectionate, but Maine Coons tend to be more independent and curious. They enjoy exploring their environment and are typically more vocal and active. Ragdolls are more subdued, preferring to stay near their human companions and engage in less intense play.

Ragdoll vs. Persian: Persians are known for their luxurious coats and calm nature, but they tend to be more reserved and less interactive than Ragdolls. Grooming a Persian requires daily attention due to their dense undercoat, whereas Ragdolls have a low-maintenance, non-matting coat.

Ragdoll vs. Siamese: Siamese cats are high-energy, very vocal, and demand constant interaction. They form strong bonds but can be overwhelming for some owners. Ragdolls, by contrast, are quieter, gentler, and more relaxed in their engagement.

Ragdoll vs. British Shorthair: British Shorthairs are calm and reserved, often enjoying alone time. They form bonds but are not as cuddly or physically affectionate. Ragdolls are more openly loving, consistently seeking human contact and attention.

Ragdoll vs. Bengal: Bengals are active, playful, and highly energetic. They thrive in environments that offer physical and mental challenges. While Ragdolls also enjoy play, they require a much calmer lifestyle and less intense stimulation.

Key Takeaway: Compared to other breeds, Ragdolls offer a rare mix of size, serenity, and sociability, making them a favorite among those seeking a low-stress, emotionally bonded companion.

3.3 Are Ragdolls the Right Choice for You?

Choosing the right cat breed is about more than just aesthetics—it's about compatibility with your lifestyle, expectations, and ability to meet the animal's emotional and physical needs. Ragdolls make wonderful pets for a wide range of households, but it's important to evaluate if they align with your routine and personality.

Ideal Owners:

- Families with children looking for a gentle, tolerant cat

- Singles or couples seeking a companion animal

- Seniors who want a calm and comforting presence

- Multi-pet households (especially with other gentle pets)

Lifestyle Compatibility:

- Ragdolls are indoor-only cats. They should not be allowed to roam freely outdoors due to their trusting nature, which makes them vulnerable to accidents and predators.

- They thrive on routine and human interaction. If you're away from home frequently, consider adopting a second pet or using enrichment tools like cat trees, window perches, or interactive feeders.

- These cats are relatively quiet and not prone to disruptive behavior, making them ideal for apartment living.

Care Requirements:

- Moderate grooming needs (brushing 2–3 times a week)

- Balanced diet and regular vet care

- Daily interaction and enrichment activities

Time and Emotional Investment:

- Ragdolls need emotional availability. They form deep bonds and don't do well with neglect.

- They do best when they can share time with their humans throughout the day, including mornings and evenings.

- They should not be treated as decorative pets — they are full-fledged emotional companions.

Not Ideal For:

- Households with aggressive pets

- Owners seeking a fully independent, low-interaction pet

- Individuals who travel frequently without pet care arrangements

Key Takeaway: Ragdolls are well-suited for owners who desire companionship and are prepared to invest time in bonding, grooming, and daily interaction. They are not low-maintenance or aloof; they are emotionally engaged members of the family.

3.4 Common Misunderstandings About Their Behavior

Because of their unique personality traits, Ragdolls are often misunderstood—even by experienced cat owners. This section clears up some of the most common misconceptions that may impact how Ragdolls are treated or perceived.

"They're Lazy"

Ragdolls aren't lazy—they're simply more relaxed than other breeds. While they love to lounge and cuddle, they still need stimulation and enjoy playtime. Owners should not mistake their calm nature for a lack of need for exercise or attention.

"They Never Scratch or Bite"

While Ragdolls are among the most tolerant cats, they are still animals with boundaries. Like any cat, they may scratch or bite if frightened, in pain, or pushed too far. Teaching children and guests to handle them gently is key.

"They Can Be Left Alone for Long Periods"

Ragdolls crave companionship. Leaving them alone for extended hours can lead to boredom, stress, or even depression. Providing enrichment or a companion pet helps mitigate this.

"They Don't Need Socialization"

Even though Ragdolls are naturally friendly, early socialization is crucial. Kittens that are handled regularly and introduced to various people, pets, and environments tend to grow into more confident, well-adjusted adults.

"They're Always Friendly"

Most Ragdolls are very friendly, but individual personality varies. Some may be shy or take time to warm up to new people or surroundings. Patience, consistency, and positive reinforcement are essential.

"They Don't Need to Play"

Even calm cats need stimulation. Mental engagement and physical play are vital to avoid boredom-related behaviors like over-grooming or excessive sleeping. Invest in toys that mimic prey, use puzzle feeders, or dedicate 15–30 minutes per day to interactive play.

"They're Good for Allergies"

Though Ragdolls produce less Fel d 1 protein compared to some breeds, they are not hypoallergenic. People with moderate to severe cat allergies should test their reactions before adopting.

Key Takeaway: Understanding your Ragdoll's behavior helps avoid misinterpretation and ensures better communication, bonding, and long-term satisfaction for both cat and owner. A well-understood Ragdoll is a happy, thriving companion.

CHAPTER 4

Preparing Your Home for a Ragdoll

4.1 Safe Indoor Environment Setup

Ragdolls are an indoor-only breed. Their trusting, relaxed nature means they are especially vulnerable to the dangers of the outside world, including cars, aggressive animals, toxins, and theft. Creating a secure and engaging indoor environment is essential for their safety and well-being.

Start with a Quiet Base Room:

When you first bring your Ragdoll home, select a quiet, enclosed room where they can feel secure. Include food, water, a litter box, a soft bed, scratching posts, and a few toys. This space will act as a transitional sanctuary until they become comfortable enough to explore the rest of the home.

Gradually Expand Access:

Allow your cat to explore new areas slowly, one room at a time. Supervise them at first to ensure there are no hazards like open windows, dangling cords, or sharp objects. Never force exploration; let your Ragdoll lead the pace.

Minimize Loud Noises and Sudden Movements:

Avoid running vacuums, loud music, or startling activity in your cat's environment, especially during the early days. Ragdolls are gentle and may become easily stressed by chaotic surroundings.

Designate Safe Zones:

Create accessible safe zones like cozy corners, cubby beds, or cat tents where your Ragdoll can retreat when overwhelmed. These spaces should be respected as their personal territory.

Use Window Perches:

Though indoor-only, Ragdolls love watching the world outside. A soft perch near a secure window gives them mental stimulation without the risks of outdoor exposure.

Key Takeaway: Building a safe, calm, and controlled environment helps your Ragdoll adapt smoothly to their new home while staying protected from indoor hazards.

4.2 Essential Living Spaces for Cats

To thrive, Ragdolls need designated areas that fulfill their basic needs for sleep, play, hygiene, and relaxation. Thoughtful setup promotes healthy habits and strengthens your bond with your new feline family member.

Sleeping Areas:

Ragdolls love comfort. Invest in plush beds, window hammocks, or soft blankets placed in quiet corners. Some may prefer being near their humans, such as on the bed or a couch.

Litter Box Placement:

Choose a low-traffic, private location for the litter box. Avoid placing it near food or water bowls. Ensure it's easy to access and cleaned daily—Ragdolls are clean cats and may refuse a dirty box.

Feeding Stations:

Dedicate a consistent spot for meals. Use ceramic or stainless-steel bowls placed on a washable mat. Place water in a separate area, ideally a few feet away from food to encourage hydration.

Scratching and Climbing:

Provide vertical and horizontal scratching surfaces. While Ragdolls aren't big climbers, they still enjoy scratching posts and small cat trees. Sisal, cardboard, or carpeted posts are effective options.

Play and Enrichment:

Toys should be varied and rotated to maintain interest. Use soft balls, feather wands, treat puzzles, and interactive play to stimulate their hunting instincts. Include a daily play routine to keep your Ragdoll engaged.

Key Takeaway: Setting up essential living zones ensures your Ragdoll has comfort, stimulation, and hygiene support in every area of the home.

4.3 Cat-Proofing Your Home

Ragdolls are curious and trusting, which means owners must take proactive steps to remove hazards from the living space. Cat-proofing prevents injury and keeps both your pet and belongings safe.

Secure Hazardous Items:

Keep medications, cleaning agents, chemicals, and toxic plants out of reach. Even natural substances like essential oils can be harmful to cats.

Block Unsafe Spaces:

Prevent access to tight spots behind appliances or inside furniture where a cat might get stuck. Use childproof locks on cabinets, and plug unused electrical outlets.

Check Windows and Screens:

Ensure all windows have secure screens. Ragdolls may lean or press against screens, and a weak or unlatched window could lead to dangerous falls.

Tie Up Cords and Curtains:

Blind cords and dangling fabric pose a strangulation risk. Tuck cords away or use cord winders and anchors to secure them.

Protect Furniture and Wires:

Use double-sided tape or citrus sprays on furniture to deter scratching. Cover electrical wires with cord protectors to prevent chewing.

Fireplace and Kitchen Safety:

Keep your cat out of the kitchen while cooking, and never leave open flames or hot surfaces unattended. Ragdolls are not fearful of fire or hot stoves and can easily be burned.

Key Takeaway: A properly cat-proofed home prevents accidents and ensures that your curious, trusting Ragdoll stays safe and healthy.

4.4 Creating a Stress-Free Transition

Bringing a new cat home is a major adjustment, especially for a sensitive breed like the Ragdoll. A smooth transition requires patience, routine, and a focus on emotional comfort.

First 24 Hours:

Keep your Ragdoll confined to the safe base room. Limit visitors, and allow them to approach you at their own pace. Speak softly, avoid picking them up too soon, and offer treats or toys to create positive associations.

Establish Routine:

Cats thrive on routine. Feed your Ragdoll at the same times daily, maintain consistent cleaning of the litter box, and dedicate a regular time for play and grooming. Predictability builds trust.

Introduce Slowly to Family and Pets:

Let your Ragdoll initiate contact with new people. For existing pets, use gradual scent introductions followed by short, supervised face-to-face interactions. Never force introductions.

Watch for Signs of Stress:

Look for signals like hiding, excessive grooming, loss of appetite, or vocalization. Provide comfort and reassurance without overwhelming them. Consider Feliway diffusers to ease anxiety.

Positive Reinforcement:

Reward desired behaviors with soft praise, petting, and treats. This builds confidence and helps your Ragdoll feel secure in their new environment.

Build Trust Gradually:

Use gentle touch, slow movements, and eye contact (including slow blinking) to establish emotional connection. Sit near your cat without demanding attention, allowing them to initiate.

Key Takeaway: A peaceful transition is built on patience, predictability, and empathy—your Ragdoll's sense of security begins with you.

CHAPTER 5

Feeding Basics

5.1 Understanding Nutritional Needs

Ragdoll cats, like all felines, are obligate carnivores. This means their bodies are biologically designed to thrive on a diet rich in animal protein and fats. Unlike omnivores, they have specific nutritional requirements that must be met through appropriate food sources. Their diet should support muscle development, coat health, energy levels, and organ function.

Key nutritional components include:

- **High-quality animal protein:** Vital for muscle repair and overall body function.

- **Animal fats:** Provide energy and support coat and skin health.

- **Taurine:** An essential amino acid for heart and eye health, found naturally in meat.

- **Vitamins and minerals:** Necessary for immune support, metabolic function, and cellular repair.

- **Water:** Critical for kidney health, especially in cats who consume dry food.

Ragdolls also tend to be larger and less active than some other breeds, which makes portion control and balanced nutrients even more important. Overfeeding can easily lead to weight gain and health issues like joint stress or diabetes.

Age-specific nutritional needs:

- **Kittens:** Require high-protein, calorie-dense meals to support growth and development.

- **Adults:** Need maintenance nutrition with controlled calories and balanced fats.

- **Seniors:** Benefit from lower-calorie meals with joint and digestive support.

Key Takeaway: Understanding the specific dietary needs of Ragdolls is essential for maintaining their health, energy, and longevity.

5.2 Choosing Balanced Meals

Selecting the right meals for your Ragdoll involves more than reading the label. Since commercial brand names and products are excluded from this book, the focus here is on the general qualities and composition of a healthy feline diet.

Types of feeding options:

- **Wet (Canned) Food:** High in moisture, helpful for hydration and often more palatable.

- **Dry (Kibble) Food:** Convenient, shelf-stable, and good for dental texture but may lack hydration.

- **Raw Diet:** Closest to the cat's ancestral diet, but must be well-balanced, hygienically prepared, and guided by veterinary advice.

- **Homemade Cooked Meals:** Allows full control of ingredients but requires proper nutrient supplementation and balance.

Qualities of a balanced meal:

- Contains a named animal protein as the first ingredient (e.g., chicken, turkey, beef).

- Free from artificial preservatives, colorants, or excessive fillers like corn, soy, or wheat.

- Includes organ meats and essential fatty acids like omega-3s.

- Matches life stage (kitten, adult, senior) and is appropriate for the cat's weight and health condition.

Feeding considerations for Ragdolls:

- Due to their larger size, Ragdolls may require slightly more food than average-sized cats—but not excessively so.

- They are prone to gaining weight if overfed, so measured portions are crucial.

- Always provide fresh, clean water, ideally in a shallow, wide dish to encourage drinking.

Key Takeaway: The best meals for Ragdolls are meat-based, moisture-rich, and free of unnecessary fillers or additives, chosen with life stage and body condition in mind.

5.3 Feeding Schedules and Portions

Consistency in feeding routines helps Ragdolls stay healthy, maintain an ideal weight, and avoid behavioral issues related to hunger or boredom. Structured meal times also support digestion and prevent overeating.

Kittens (up to 12 months):

- Feed 3–4 small meals daily.

- Offer calorie-dense, high-protein meals for growth.

Adults (1–7 years):

- Feed 2 meals per day.

- Monitor body condition score (BCS) regularly.

- Adjust portions based on activity level and metabolism.

Seniors (7+ years):

- Feed 2–3 smaller, easier-to-digest meals.

- Focus on joint and digestive support.

Portion guidelines:

- Always use a measuring cup or kitchen scale for accuracy.

- Start with the recommended amount for your cat's weight and adjust over time.

- Monitor for changes in weight, appetite, and stool quality as signs of dietary needs.

Scheduled feeding vs. free feeding:

- **Scheduled feeding** (recommended): Promotes portion control, digestion, and behavioral structure.

- **Free feeding** (not ideal): Can lead to overeating and obesity.

Water intake:

- Encourage hydration by providing multiple water bowls or pet fountains.

- Ragdolls, like many cats, have a low thirst drive, so moisture in food is especially helpful.

Key Takeaway: Feeding your Ragdoll at consistent times with proper portion control encourages healthy eating habits and weight maintenance.

5.4 Avoiding Common Feeding Mistakes

Even well-meaning owners can make mistakes when it comes to feeding. This section highlights the most common pitfalls and how to avoid them to ensure your Ragdoll remains healthy and happy.

Overfeeding:

- One of the most common issues. Even slight overfeeding, repeated daily, can lead to weight gain.

- Use feeding guidelines based on weight and monitor for gradual increases.

Feeding only dry food:

- A diet of exclusively dry food may lack adequate moisture, increasing the risk of urinary tract issues.

- Supplement with wet meals or add water to kibble if necessary.

Ignoring food allergies or sensitivities:

- Some Ragdolls may react poorly to certain proteins or grains.

- Watch for signs like vomiting, diarrhea, itching, or excessive grooming.

Feeding from the table:

- Human food is often too rich or toxic for cats.

- Avoid giving scraps or cooked bones, and be cautious of onions, garlic, and dairy.

Sudden food changes:

- Transition gradually over 7–10 days when switching foods.

- Sudden changes can cause gastrointestinal upset.

Using the wrong dish type:

- Plastic bowls can harbor bacteria and cause chin acne.

- Use stainless steel or ceramic dishes, cleaned daily.

Neglecting dental health:

- Food texture does not replace dental care.

- Include brushing, dental toys, or approved oral hygiene options.

Key Takeaway: A thoughtful feeding approach avoids common mistakes and supports your Ragdoll's long-term health and happiness.

CHAPTER 6

Housing and Space Needs

6.1 Indoor Living for Ragdolls

Ragdoll cats are best suited for an indoor lifestyle. Their laid-back, trusting nature makes them ill-equipped for the dangers of outdoor living. They lack the heightened wariness that helps other breeds avoid predators, traffic, and toxins. Therefore, setting up a comfortable, stimulating indoor environment is essential for their physical safety and emotional well-being.

Benefits of Indoor Living:

- **Safety:** Protection from cars, predators, parasites, and disease.

- **Health:** Reduced risk of injuries and exposure to contagious feline illnesses.

- **Lifespan:** Indoor cats tend to live longer, healthier lives.

Indoor Environment Considerations:

- **Space:** Ragdolls don't require vast amounts of space, but they do need areas to stretch, explore, and rest.

- **Vertical Options:** While not avid climbers, they enjoy soft cat trees and window perches.

- **Lighting:** Sun-filled areas are appreciated for warmth and napping.

Daily Activities:

- Offer gentle playtime each day.

- Rotate toys to maintain interest.

- Create viewing spots to observe the outside world safely.

Quiet and Calm:

Ragdolls thrive in peaceful households. Avoid constant loud noise, sudden movements, or overcrowded environments that might stress them.

Key Takeaway: An indoor-only lifestyle protects your Ragdoll from harm while offering the comfort and stimulation they need for a long, content life.

6.2 Ideal Sleeping Areas and Rest Zones

Ragdolls love to sleep—and they sleep a lot. On average, adult cats can sleep up to 16 hours a day, with kittens and seniors sleeping even more. Providing multiple cozy and accessible resting areas throughout the home is important.

Preferred Sleeping Styles:

- Ragdolls enjoy soft, supportive surfaces like fleece, plush bedding, or warm blankets.

- Some prefer enclosed beds (caves or tunnels), while others like open, elevated perches.

Rest Zones You Can Create:

- **Window Perches:** Great for naps in the sun.

- **Heated Pads or Beds:** Especially appreciated in cooler climates or for senior cats.

- **Bedroom Corners:** Ragdolls often enjoy sleeping near or on the bed with their owners.

- **Quiet Hideaways:** Useful for shy cats or those that need a break from stimulation.

Number of Beds:

Offer at least 2–3 comfortable sleeping spots in different parts of the home, especially in high-traffic and low-traffic zones.

Scented Comfort:

Place items with your scent—like a worn T-shirt—near sleeping areas to help your cat feel emotionally secure.

Key Takeaway: Ragdolls need multiple soft, peaceful sleeping spots that offer warmth, security, and variety based on their mood and environment.

6.3 Multi-Pet Households Considerations

Ragdolls are known for their tolerant and friendly behavior, making them a good choice for households with other pets.

However, introductions should be made with care, and ongoing interactions should be managed to prevent stress or conflict.

When Introducing to Other Cats:

- Use scent exchange (blankets or toys).

- Start with separated rooms.

- Allow slow, supervised introductions over several days to weeks.

With Dogs:

- Choose calm, cat-friendly dog breeds.

- Keep the dog on a leash for initial meetings.

- Never allow the dog to chase or corner the cat.

With Small Pets (birds, rodents, etc.):

- Supervise all interactions.

- Keep small animals in secure enclosures.

- Ragdolls have a hunting instinct, though milder than other breeds.

Tips for Harmony:

- Provide enough space, litter boxes, and resting areas for each pet.

- Ensure each animal has escape or retreat options.

- Don't force interactions—let relationships form naturally.

Monitor for Stress Signs:

- Aggression, hiding, food avoidance, or excessive grooming can indicate discomfort.

- Maintain routine and reduce triggers when tension arises.

Key Takeaway: With patience, structure, and supervision, Ragdolls can thrive in homes shared with other pets.

6.4 Outdoor Access and Safety Concerns

While Ragdolls are indoor cats, some owners choose to provide safe, supervised outdoor time through secure setups. If done thoughtfully, this can enrich their environment without compromising safety.

Why Ragdolls Shouldn't Roam Freely:

- Highly trusting and unlikely to defend themselves.

- Vulnerable to theft due to their rarity and beauty.

- Easily injured or killed by cars, dogs, or wildlife.

- Susceptible to parasites, poisons, and contagious diseases.

Safe Outdoor Alternatives:

- **Catios:** Enclosed outdoor patios built specifically for cats. Allow fresh air and safe stimulation.

- **Harness and Leash Training:** Train your Ragdoll gradually using a padded harness. Supervise all outdoor walks.

- **Window Perches and Bird Feeders:** Great indoor-outdoor connection that satisfies curiosity and hunting instincts.

- **Supervised Yard Time:** Fully fenced, escape-proof yards with human supervision can be safe if carefully monitored.

Outdoor Enrichment Without Risk:

- Place bird feeders outside secure windows.

- Use scent and sound enrichment (cat-safe herbs, recorded bird sounds).

Key Takeaway: Free-roaming is unsafe for Ragdolls, but creative indoor-outdoor solutions can enrich their lives without exposing them to danger.

CHAPTER 7

Grooming and Coat Care

7.1 Brushing and Detangling Ragdoll Fur

Ragdolls have semi-long, silky fur that lacks the dense undercoat found in other long-haired breeds. This makes their coat less prone to matting, but it still requires regular grooming to keep it clean, soft, and free of tangles.

Brushing Frequency:

- Brush your Ragdoll 2–3 times a week.

- Increase to daily brushing during seasonal shedding periods (spring and fall).

Tools You'll Need:

- Wide-tooth comb: For gentle detangling.

- Slicker brush: For removing loose fur.

- Fine-tooth comb: For face and delicate areas.

Detangling Tips:

- Always brush in the direction of hair growth.

- Use your fingers to gently loosen knots before using tools.

- Be extra gentle around sensitive areas like the belly, armpits, and behind the ears.

Grooming Bonding Time:

Turn brushing into a bonding session by speaking softly, using slow movements, and rewarding with gentle praise or treats afterward.

Benefits of Regular Brushing:

- Reduces shedding and hairballs.

- Prevents mats and tangles.

- Promotes circulation and a healthy coat.

- Builds trust and comfort between you and your cat.

Key Takeaway: Routine brushing not only maintains coat health but also strengthens the emotional bond with your Ragdoll.

7.2 Bathing: When and How

While cats are generally self-cleaning, Ragdolls sometimes benefit from occasional baths—especially if they become soiled, oily, or develop allergens on their coat.

When to Bathe:

- Once every few months, if needed.

- More frequently if your cat has allergies or gets into messes.

- If your Ragdoll has long fur prone to oil build-up.

Bathing Preparation:

- Trim nails beforehand to prevent scratching.

- Use a non-slip mat in the sink or tub.

- Brush out tangles before bathing.

Bathing Steps:

1. Fill the tub/sink with lukewarm water (2–3 inches deep).

2. Wet your cat gently with a cup or sprayer, avoiding the face.

3. Use a mild, cat-safe shampoo.

4. Lather gently from neck to tail.

5. Rinse thoroughly to remove all soap.

6. Dry with a soft towel; avoid using a loud blow dryer unless your cat is accustomed to it.

Aftercare:

- Keep your Ragdoll in a warm room until fully dry.

- Offer treats and positive reinforcement.

Key Takeaway: Occasional, gentle bathing supports coat hygiene and allergen control—when done calmly and correctly.

7.3 Nail Care and Paw Maintenance

Ragdolls are indoor cats and typically don't wear down their nails naturally, so regular trimming is essential for their comfort and health.

How Often to Trim Nails:

- Every 2–3 weeks.

Nail Trimming Tips:

- Use cat-specific nail clippers.

- Trim only the sharp tip; avoid cutting into the pink "quick."

- Work on one paw at a time, offering praise between each.

If Your Cat Resists:

- Gently wrap them in a towel with one paw exposed.

- Trim during sleepy times or after a meal.

- Start by handling paws daily to build tolerance.

Paw Health Checks:

- Inspect for debris, redness, or swelling.

- Clean gently with a damp cloth if dirty.

- Keep litter clean to prevent infection.

Claw Covers:

Some owners use soft vinyl claw caps to prevent scratching—consult with a vet first.

Never Declaw:

Declawing is inhumane and leads to long-term pain and behavioral issues.

Key Takeaway: Nail trimming and paw care protect your Ragdoll's comfort and prevent household damage.

7.4 Eye, Ear, and Dental Hygiene

Good grooming doesn't stop at fur and nails—your Ragdoll's eyes, ears, and teeth also require regular attention.

Eye Care:

- Use a soft, damp cloth to wipe away any discharge.

- Always wipe outward from the eye, using a separate section of the cloth for each eye.

- Excessive tearing may indicate allergies or a health issue—consult a vet.

Ear Cleaning:

- Check weekly for wax build-up, dirt, or odor.
- Use a vet-approved ear-cleaning solution with a cotton ball.
- Never insert cotton swabs deep into the ear canal.

Signs of Infection:

- Redness, foul odor, scratching at ears, head shaking.

Dental Hygiene:

- Brush teeth 2–3 times per week with a feline-safe toothpaste.
- Use finger brushes or cat-specific toothbrushes.
- Offer dental toys or treats that support oral health (brand-free).

Watch for Dental Issues:

- Bad breath, drooling, difficulty eating, red gums.

Key Takeaway: Grooming should include routine care of your Ragdoll's eyes, ears, and teeth to prevent disease and promote long-term health.

CHAPTER 8

Training Fundamentals

8.1 Why Training Matters for Cats

Contrary to popular belief, cats can be trained just like dogs—but with a different approach. While cats are independent by nature, Ragdolls are notably more responsive to human interaction and routine. Training your Ragdoll not only enhances behavior but also deepens your bond and ensures a safer, more harmonious home.

Benefits of Training Ragdolls:

- **Behavioral structure:** Sets boundaries and encourages calm routines.

- **Safety:** Prevents dangerous behaviors like jumping on stoves or chewing cords.

- **Mental stimulation:** Engages their intelligence and reduces boredom.

- **Stress reduction:** Predictable environments and expectations help prevent anxiety.

- **Social development:** Encourages gentle interactions with humans and other pets.

Common Training Goals for Ragdolls:

- Litter box use

- Scratching post use

- Staying off counters or forbidden areas

- Gentle play without biting or scratching

- Coming when called or responding to names

Key Takeaway: Training is not about controlling your cat—it's about communication, cooperation, and care.

8.2 Positive Reinforcement Techniques

Ragdolls respond best to gentle, consistent training that rewards good behavior rather than punishing bad habits.

Positive reinforcement builds trust and teaches your cat that desired behaviors lead to rewards.

Basic Training Principles:

- **Reward the behavior you want.**

- **Ignore or redirect the behavior you don't.**

- **Be consistent and patient.**

- **Always end training sessions on a positive note.**

Reward Ideas (non-commercial):

- Extra petting or brushing

- Verbal praise in a soft, affectionate tone

- Small pieces of cooked chicken or freeze-dried meat (natural treats)

- Playtime with favorite toy

Timing is Critical:

- Give the reward immediately after the desired behavior.

- Delay reduces the connection between action and reward.

Short Sessions Work Best:

- Keep training sessions to 5–10 minutes.

- Repeat 1–2 times per day.

- Avoid training when your cat is tired, stressed, or hungry.

Clicker Training:

- Some owners use a soft clicker sound to mark good behavior before rewarding.

- This method can help with precision training and trick teaching.

Key Takeaway: Ragdolls thrive on kindness and attention— positive reinforcement encourages long-term learning and trust.

8.3 Teaching Boundaries and Rules

Teaching your Ragdoll boundaries early on helps prevent frustration and maintains a peaceful home. Gentle discipline paired with consistency is key.

Common Boundary Lessons:

- **No jumping on kitchen counters.**

- **Using designated scratching posts.**

- **No biting or scratching during play.**

- **Staying off furniture, if desired.**

Effective Boundary Techniques:

- **Redirect:** Guide your cat to a preferred behavior (e.g., place them on a scratching post).

- **Remove temptation:** Use double-sided tape or foil on forbidden areas temporarily.

- **Ignore attention-seeking misbehavior.**

- **Reward compliance:** Praise your cat when they obey the rule.

Do Not:

- Shout or physically punish your cat.

- Use spray bottles or other scare-based tactics—these damage trust.

- Punish after the fact—cats won't understand.

Consistency is Everything:

- Make sure all family members follow the same rules.

- Reinforce boundaries regularly.

- Be patient—it can take weeks for habits to form.

Key Takeaway: Clear, gentle boundaries help your Ragdoll understand household expectations without fear or confusion.

8.4 Litter Training and Success Tips

Most Ragdolls are naturally inclined to use a litter box, especially when introduced to one as kittens. Still, setting them up for success with proper technique is essential.

Choosing a Litter Box:

- Size: Big enough for your cat to enter, turn, and dig comfortably.

- Type: Open boxes allow better airflow; covered ones offer privacy.

- Number: One box per cat, plus one extra if you have multiple cats.

Litter Considerations:

- Use unscented, clumping litter for easy cleaning.

- Avoid harsh chemicals or fragrances.

- Scoop daily; clean the box weekly with mild soap and warm water.

Location Matters:

- Place in quiet, accessible areas.

- Avoid placing near food or water.

- Ensure it's easy for kittens and older cats to reach.

Training Steps:

1. Show your cat the box and gently place them inside after meals or naps.

2. Reward or praise when they use it.

3. If accidents occur, clean the area thoroughly and try moving the box to that spot.

4. Never punish for accidents—this creates fear.

Signs of Litter Problems:

- Avoidance may signal stress, illness, or litter preference.

- Monitor for signs of urinary issues, like straining or frequent trips.

Key Takeaway: With the right setup and patience, litter training is usually smooth—especially with intuitive, clean-loving breeds like Ragdolls.

CHAPTER 9

Socialization and Bonding

9.1 Helping Your Cat Adapt to New People

Ragdolls are naturally affectionate and social, but even they can feel shy or overwhelmed when meeting unfamiliar people. Proper socialization ensures your cat feels safe, secure, and confident around guests, family members, and new caretakers.

Initial Introductions:

- Allow your cat to approach at their own pace.

- Avoid forcing contact—let them sniff and observe.

- Keep initial visits short and calm.

- Encourage guests to sit quietly and extend a hand for sniffing.

Reading Body Language:

- Signs of comfort: slow blinking, soft body, purring.

- Signs of fear: flattened ears, hiding, tail tucked, hissing.

Tips for Children:

- Teach kids how to pet gently and avoid tail pulling.

- Always supervise interactions.

- Provide the cat with an escape route or quiet space.

Gradual Exposure:

- Start with quiet visitors.

- Introduce people during feeding time or play to associate new faces with positive experiences.

Consistency and Routine:

- Ragdolls are routine-driven. Frequent, low-stress interactions build familiarity.

Key Takeaway: Positive and patient introductions help Ragdolls become confident around new people and environments.

9.2 Bonding Activities for Trust Building

Bonding with your Ragdoll builds the foundation for lifelong companionship. These gentle cats respond deeply to love, attention, and respectful handling.

Daily Bonding Rituals:

- **Grooming sessions:** Brushing not only keeps the coat healthy but reinforces trust.

- **Interactive play:** Wand toys and soft chase games strengthen your connection.

- **Lap time or cuddling:** Let your cat choose when to snuggle—don't force contact.

- **Verbal bonding:** Speak softly and call them by name consistently.

Respecting Boundaries:

- Let your cat initiate closeness.

- Avoid overhandling or picking up if they resist.

- Never disturb a sleeping or hiding cat—this can create stress.

Signs of Deep Bonding:

- Following you around the house

- Head butting or rubbing

- Slow blinks and purring

- Bringing you "gifts" (toys, objects)

Engage Their Senses:

- Use scent (your clothing, bedding) to provide comfort.

- Try clicker training as a fun, reward-based bonding tool.

Key Takeaway: Bonding takes time, patience, and mutual respect—nurture your Ragdoll with consistency and love.

9.3 Introducing Ragdolls to Other Pets

Ragdolls tend to get along well with other animals when introduced properly. Whether you're bringing a Ragdoll into a multi-pet home or adding a new animal to your existing Ragdoll's space, gradual and calm introductions are key.

Introduction Timeline:

- **Week 1:** Keep animals in separate rooms. Exchange scents using blankets, toys, or bedding.

- **Week 2:** Begin short, supervised visual introductions through a baby gate or cracked door.

- **Week 3:** Allow limited supervised contact. End sessions before signs of stress.

Tips for Introducing to Other Cats:

- Provide separate litter boxes, food bowls, and resting spots.

- Monitor for territorial behavior.

- Use Feliway diffusers to reduce stress.

Tips for Introducing to Dogs:

- Choose calm, obedient dog breeds.

- Keep the dog on a leash during initial meetings.

- Train the dog to ignore the cat and reward calm behavior.

Monitor Interactions:

- Watch for warning signs like growling, raised fur, or retreating.

- Allow the Ragdoll to control the pace.

Avoid Forced Contact:

- Give each pet their own space to retreat.

- Maintain routines to minimize jealousy or disruption.

Key Takeaway: Thoughtful, patient introductions create peaceful multi-pet households where your Ragdoll can thrive socially.

9.4 Creating a Loving Daily Routine

Ragdolls flourish in structured, loving environments. A consistent daily routine helps them feel secure and strengthens the human-animal bond.

Sample Daily Routine:

- **Morning:**

    - Soft greeting and petting

    - Fresh food and water

    - Quick grooming session

    - Window time or enrichment activity

- **Midday:**

    - Interactive play for mental and physical stimulation

    - Quiet rest time

    - Occasional bonding session (lap time, brushing)

- **Evening:**

    - Main meal and litter box cleaning

    - One-on-one play

    - Gentle petting and winding down

Weekend Variations:

- Extended cuddle or grooming sessions

- More time for outdoor-safe activities like harness walks or catio time

Routine Enhancers:

- Feed meals at the same time each day.

- Speak to your cat consistently.

- Keep litter box, feeding, and sleeping areas clean and in the same locations.

Routines Build Trust:

- Predictability reduces stress and behavioral issues.

- Routine encourages independence while maintaining emotional closeness.

Key Takeaway: A loving daily routine meets your Ragdoll's physical, emotional, and social needs—creating a happy, confident companion.

CHAPTER 10

Understanding Ragdoll Behavior

10.1 Natural Instincts of Ragdolls

While Ragdolls are famously known for their relaxed and docile temperament, they still retain many of the natural feline instincts shared by all cats. Understanding these underlying behaviors helps pet parents provide an environment that supports both comfort and stimulation.

Hunting Instincts:

- Though domesticated, Ragdolls retain a playful hunting drive.

- They often enjoy toys that mimic prey—like feather wands, moving objects, or hide-and-seek puzzles.

- This instinct is strongest in the early morning or evening (crepuscular activity).

Territorial Behavior:

- Ragdolls may stake quiet claim to certain napping spots, scratching posts, or windowsills.

- Though non-aggressive, they enjoy having familiar zones.

- Introducing new pets or rearranging furniture may disrupt their sense of control.

Curiosity and Exploration:

- Ragdolls are gentle explorers. They'll follow their humans, check open doors, and sniff out new objects.

- While they're less mischievous than some breeds, they still appreciate environmental variety.

Social Dependency:

- This breed thrives on companionship. Being alone for long hours can cause boredom or stress.

- They often wait at doors, follow their owners, or vocalize when seeking company.

Key Takeaway: Ragdolls are both calm and instinctual—understanding their subtle wild roots helps you enrich their daily lives.

10.2 Recognizing Body Language

Feline communication is subtle, but reading your Ragdoll's body language can prevent misunderstandings and strengthen your relationship.

Tail Positions:

- **Upright, relaxed:** Confidence and happiness.

- **Tucked or puffed:** Fear or stress.

- **Slow swish:** Focus or irritation.

Ears:

- **Forward:** Alert and curious.

- **Sideways or back:** Anxiety or overstimulation.

- **Flat against head:** Defensive or scared.

Eyes:

- **Slow blink:** Trust and affection.

- **Dilated pupils:** Excitement, fear, or hunting mode.

- **Avoiding eye contact:** Submissiveness or calm disengagement.

Body Posture:

- **Lying belly up:** Extreme trust.

- **Arched back with fur raised:** Fear or startle response.

- **Crouching:** Readiness to pounce or nervousness.

Vocal Cues:

- Ragdolls are typically quiet but may coo, chirp, or softly meow.

- Increased vocalization can signal neediness, discomfort, or illness.

Key Takeaway: Reading your cat's body cues allows you to respond appropriately to their needs and feelings.

10.3 Affectionate vs. Independent Behaviors

One of the Ragdoll's defining characteristics is their balanced temperament. They are affectionate but not clingy, sociable but not overly demanding.

Affectionate Signs:

- Following you from room to room

- Sleeping on or beside you

- Purring during petting or brushing

- Head butting or cheek rubbing

- Bringing toys to initiate play

Independent Moments:

- Seeking quiet alone time in a cozy hideaway

- Observing from a distance without engaging

- Grooming themselves or lounging without interaction

What This Balance Means for Owners:

- Ragdolls enjoy companionship but won't constantly demand attention.

- They'll initiate cuddles but also respect personal space.

- Encouraging independence supports their confidence and reduces stress when you're away.

Avoid Misinterpreting Independence:

- Not all solitude means something's wrong.

- Changes in affectionate behavior can signal illness, stress, or environmental changes.

Key Takeaway: Understanding when your Ragdoll needs affection versus space leads to mutual respect and a better bond.

10.4 Signs of Stress or Discomfort

Even with their gentle demeanor, Ragdolls can experience anxiety, fear, or overstimulation. Knowing the warning signs helps prevent behavioral issues and supports overall wellness.

Common Stress Triggers:

- Loud noises or frequent visitors

- Travel, vet visits, or moving homes

- Unfamiliar pets or people

- Poor litter box hygiene or changes

Behavioral Indicators of Stress:

- Hiding more than usual

- Overgrooming or fur loss

- Reduced appetite or sudden weight change

- Aggression or sudden fearfulness

- Inappropriate elimination (outside litter box)

Physical Symptoms of Discomfort:

- Frequent vomiting or diarrhea

- Excessive panting or drooling

- Constant scratching or biting at skin

- Lethargy or restlessness

How to Help:

- Create quiet spaces or elevated resting areas.

- Use gentle, calming tones.

- Maintain a consistent routine.

- Provide enrichment (scratching posts, puzzle feeders).

- Consult a veterinarian if symptoms persist.

Key Takeaway: Timely identification of stress signals ensures a calm, content life for your Ragdoll and prevents long-term behavioral issues.

CHAPTER 11

Play, Enrichment, and Stimulation

11.1 Importance of Play for Ragdolls

While Ragdolls are known for their calm and gentle demeanor, they still need regular play and mental stimulation to stay healthy and happy. Play is not just about physical exercise—it also nurtures their hunting instincts, prevents boredom, and strengthens the bond with their human companions.

Why Play Matters:

- **Physical exercise:** Keeps weight in check and promotes joint flexibility.

- **Mental stimulation:** Reduces stress, anxiety, and destructive behaviors.

- **Behavioral benefits:** Redirects energy and reduces aggression.

- **Emotional connection:** Creates trust and affection between cat and owner.

How Much Play Do They Need?

- **Kittens:** 3–4 short sessions (10–15 minutes each) daily.

- **Adults:** At least 1–2 active play sessions per day.

- **Seniors:** Gentle, low-impact play that stimulates without overexerting.

Best Times to Play:

- Early mornings and evenings (when cats are most alert).

- After meals or naps to prevent lazy behavior.

Key Takeaway: Daily play isn't optional—it's essential to your Ragdoll's health and happiness.

11.2 Types of Enrichment Activities

Enrichment involves providing variety in your cat's environment, encouraging natural behaviors, and preventing monotony. Ragdolls thrive when their senses are engaged regularly.

Physical Enrichment:

- Climbing trees or cat towers

- Tunnels and hideaways

- Safe spaces for exploration

Sensory Enrichment:

- Bird-watching from window perches

- Soft background music or nature sounds

- Cat grass or safe indoor plants to sniff

Mental Enrichment:

- Puzzle feeders or treat-dispensing toys

- DIY scent trails using socks or blankets

- Hiding toys for them to find

Environmental Enrichment:

- Rearranging toys or furniture occasionally

- Rotating new items into their play zone

- Setting up cardboard boxes for exploration

Key Takeaway: Variety is vital—mix up textures, activities, and spaces to keep your Ragdoll alert and engaged.

11.3 Interactive Play for Owner-Cat Bond

Interactive play is a powerful tool for strengthening your relationship with your Ragdoll. It reinforces trust, creates positive associations, and helps your cat feel more secure.

Best Tools for Interactive Play:

- Wand toys or feather sticks

- Soft balls or felt mice

- Laser pointers (always end play with a toy your cat can catch)

- Crinkle tunnels or cat-safe climbing setups

Tips for Success:

- Let your Ragdoll "win" often to build confidence.

- Vary movement patterns—mimic prey with slow and fast motions.

- Use verbal cues like your cat's name or a fun command to start play.

End Each Session Positively:

- Offer a small natural treat.

- Cuddle or groom afterward if your cat is receptive.

Scheduling Interactive Play:

- Make it part of your daily routine.

- Try to play at consistent times—cats love predictability.

Key Takeaway: You are your Ragdoll's favorite toy—interactive play deepens your bond more than solo play ever could.

11.4 Preventing Boredom and Stress

Ragdolls are intelligent and emotionally sensitive. When their mental and physical needs aren't met, they may display signs of boredom or stress. This can lead to behavioral problems, anxiety, or even health issues.

Signs of Boredom or Under-Stimulation:

- Overeating or sleeping excessively

- Destructive scratching or chewing

- Vocalizing or attention-seeking behavior

- Pacing or repetitive movements

Preventative Strategies:

- **Structured play:** Make it part of your cat's routine.

- **Rotation:** Change out toys every few days.

- **Window perches:** Provide visual enrichment.

- **Solo entertainment:** Offer toys that can be used without your involvement.

Enrichment for Times Alone:

- Puzzle feeders for treats

- Automated interactive toys (battery-operated mice)

- Hiding treats around the home

Environmental Comfort:

- Ensure your Ragdoll has quiet retreats

- Provide vertical spaces (shelves or trees)

- Avoid overcrowded or chaotic spaces

Key Takeaway: An enriched life is a happy life—daily stimulation prevents boredom and supports emotional well-being for your gentle Ragdoll.

CHAPTER 12

Health Basics

12.1 Routine Veterinary Visits

Routine veterinary care is essential to keep your Ragdoll cat in optimal health. Regular checkups allow for early detection of potential problems and provide peace of mind for pet parents.

Why Routine Vet Visits Matter:

- Establishes a health baseline for your cat

- Enables early detection of illness or discomfort

- Ensures vaccination and preventive schedules are followed

- Builds a history for future diagnoses

Recommended Schedule:

- **Kittens:** First vet visit at 6–8 weeks old, then every 3–4 weeks until 16 weeks

- **Adults (1–7 years):** Annually for exams and preventive care

- **Seniors (7+ years):** Biannual exams to monitor age-related changes

What to Expect During Visits:

- Physical examination (weight, teeth, ears, eyes, coat)

- Listening to heart and lungs

- Discussion of feeding, behavior, and overall condition

- Fecal testing, bloodwork, and dental checkups as needed

Vet Selection Tips:

- Choose a feline-friendly or fear-free certified practice

- Ensure the vet is experienced with long-haired and purebred cats

Key Takeaway: Consistent vet visits help catch issues early and foster a long, healthy life for your Ragdoll.

12.2 Vaccinations and Preventive Care

Vaccinations and parasite prevention are critical in protecting your Ragdoll against common diseases and infestations. While Ragdolls are primarily indoor cats, they still need protection.

Core Vaccinations (typically recommended):

- FVRCP: Protects against Feline Viral Rhinotracheitis, Calicivirus, and Panleukopenia

- Rabies: Required by law in many areas

Non-Core Vaccines (as advised by your vet):

- Based on lifestyle, risk, and region (e.g., FeLV for multi-cat households)

Vaccination Timeline (Typical):

- Initial vaccines during kittenhood (6–16 weeks)

- Boosters at one year

- Adult boosters every 1–3 years depending on vaccine type and vet advice

Parasite Prevention:

- Flea and tick control

- Regular deworming (based on vet recommendation)

- Heartworm prevention (in some areas)

Annual Preventive Checklist:

- Dental examination

- Weight and diet review

- Behavior and litter habits

Key Takeaway: Staying current with vaccines and parasite control is essential even for indoor Ragdolls to avoid preventable illnesses.

12.3 Common Health Issues in Ragdolls

While Ragdolls are generally healthy and robust, like all purebred cats, they are predisposed to certain genetic conditions. Awareness allows for proactive management and regular screening.

1. Hypertrophic Cardiomyopathy (HCM):

- A genetic heart disease causing thickening of the heart muscles

- May lead to heart failure if untreated

- Ask your breeder if the kitten's parents were screened

- Regular cardiac checkups recommended

2. Polycystic Kidney Disease (PKD):

- Can cause kidney cysts and eventual kidney failure

- DNA testing is available to detect risk

3. Feline Infectious Peritonitis (FIP):

- Rare but serious viral disease that affects the abdomen, chest, or nervous system

- No cure, but early diagnosis can improve comfort care

4. Obesity:

- Ragdolls are large and sedentary, making them prone to weight gain

- Monitor diet and provide exercise/enrichment

5. Urinary Tract Issues:

- Ensure adequate hydration and litter box hygiene

- Watch for straining, frequent urination, or blood in urine

6. Dental Disease:

- Regular brushing and dental checks reduce risk

Key Takeaway: Genetic screening, preventive care, and attentive monitoring help reduce the risk and impact of breed-specific health conditions.

12.4 Spotting Early Warning Signs

Knowing how to recognize early signs of illness can mean the difference between a manageable condition and a medical emergency. Ragdolls are subtle communicators, so owners must stay observant.

Physical Signs to Watch For:

- Sudden weight loss or gain

- Changes in appetite or water consumption

- Vomiting or diarrhea

- Difficulty breathing or rapid panting

- Discharge from eyes, ears, or nose

Behavioral Red Flags:

- Hiding more than usual

- Uncharacteristic aggression or irritability

- Lethargy or sudden change in activity

- Excessive grooming or fur loss

Litter Box Clues:

- Straining to urinate

- Blood in urine or feces

- Going outside the litter box

When to Call the Vet Immediately:

- Persistent vomiting or diarrhea (more than 24 hours)

- Labored breathing

- Seizures

- Collapse or unresponsiveness

Preventive Monitoring Tools:

- Keep a health journal: record weight, appetite, mood, and litter habits

- Schedule routine wellness checks

Key Takeaway: Your Ragdoll's subtle symptoms can signal major health issues—early detection saves lives and improves outcomes.

CHAPTER 14

Growth and Life Stages

14.1 Kitten Development Stages

Understanding the growth trajectory of a Ragdoll cat from birth through adulthood allows pet owners to provide age-appropriate care, anticipate behavior changes, and promote healthy development.

Neonatal Stage (0–2 Weeks):

- Eyes and ears are closed

- Kittens rely entirely on their mother for warmth, feeding, and stimulation

- Weight should steadily increase daily

Transition Stage (2–4 Weeks):

- Eyes begin to open; kittens start to hear and crawl

- Social behaviors start forming as they interact with littermates

- Introduction to very soft solid food may begin at the tail end of this stage

Socialization Stage (4–12 Weeks):

- Critical period for social learning—handling, exposure to sounds, and positive experiences are essential

- Litter training begins

- First vet visits and vaccinations

- Weaning completes around 8 weeks; kittens begin to eat on their own

Juvenile Stage (3–6 Months):

- Energetic, playful, and curious

- Rapid growth and development

- Spay/neuter is often scheduled in this period

Key Takeaway: Early life stages are foundational for health and temperament. Proper nutrition, gentle handling, and vet care ensure healthy growth and social confidence.

14.2 Caring for Adolescent Ragdolls

The adolescent stage—roughly 6 to 12 months—is often described as the feline "teenager" phase. Ragdolls grow quickly during this time and may test boundaries while developing independence.

Behavioral Changes:

- Increased energy and curiosity

- May test limits or exhibit mild rebellious behavior

- Forming long-term personality traits

Care Recommendations:

- Maintain structured routines for feeding, play, and sleep

- Continue gentle training to reinforce boundaries

- Provide mental stimulation through puzzles and varied play

- Offer social interaction to reduce loneliness

Nutritional Needs:

- Still require kitten food or high-calorie diets until fully grown (typically 12–18 months)

- Maintain portion control to prevent early weight gain

Health Considerations:

- Monitor growth and body condition regularly

- Schedule booster vaccinations

- Begin grooming habits for long-term ease

Key Takeaway: This stage sets the tone for adult behavior and habits. Patience and consistency are essential for raising a well-adjusted Ragdoll.

14.3 Adult Cat Care Essentials

Ragdolls typically reach full physical and emotional maturity between 18 months and 3 years. As adults, they settle into their unique personalities and require maintenance-focused care.

Lifestyle Characteristics:

- Ragdolls tend to be calm and affectionate

- Less hyperactive than in their adolescent stage

- Often develop strong attachment to routines and favorite resting spots

Feeding and Weight Management:

- Transition to adult-formulated food with appropriate calorie levels

- Watch for creeping weight gain due to lower energy expenditure

- Maintain hydration and dental care routines

Environmental Enrichment:

- Rotate toys to keep playtime engaging

- Provide scratching posts, perches, and observation spots

- Introduce scent enrichment (cat-safe herbs, new textures)

Health Focus:

- Annual wellness exams become more important

- Watch for early signs of disease, especially dental, joint, and weight-related issues

Key Takeaway: Adulthood is about maintaining optimal wellness and reinforcing the loving bond you've built with your Ragdoll.

14.4 Senior Cat Adjustments

Ragdolls are long-lived cats and may enjoy life well into their teens. Around 8–10 years of age, they are considered seniors, and their care needs evolve accordingly.

Physical Changes:

- Slower movement, increased sleeping

- May develop arthritis or joint stiffness

- Reduced muscle mass and slower metabolism

Dietary Adjustments:

- Lower calorie, senior-specific diets may be recommended

- Add moisture through wet food or broths to aid digestion and hydration

- Monitor appetite changes closely

Home Environment:

- Provide easy-access beds and litter boxes

- Use pet steps or ramps for elevated furniture

- Keep their environment consistent and calm

Health Monitoring:

- Semi-annual vet visits are recommended

- Dental issues, kidney disease, and hyperthyroidism become more common

- Early detection of changes in weight, behavior, or litter habits is crucial

Emotional Care:

- Gentle affection and quality time are more meaningful than ever

- Respect their desire for peace and comfort

Key Takeaway: Senior Ragdolls need tailored care, softer environments, and attentive health monitoring to enjoy a happy and graceful aging process.

CHAPTER 15

Handling and Interaction

15.1 Correct Ways to Pick Up and Hold Ragdolls

Ragdolls are famously docile and often enjoy being held—hence their name. However, proper handling is essential to ensure their comfort and safety.

How to Pick Up a Ragdoll Cat:

1. Approach calmly, speaking softly to avoid startling them.

2. Place one hand under the chest and the other supporting the hindquarters.

3. Lift gently and hold them close to your body for added security.

4. Keep their body horizontal and well-supported.

Do's and Don'ts:

- Always support their full body weight.

- Hold close to your chest to provide security.

- Don't dangle them by the front legs or lift them under the armpits.

- Avoid forcing them into uncomfortable positions.

Reading Their Comfort:

- A relaxed cat will have a soft body and may purr or rest against you.

- A tense or squirming cat may prefer to be let down— respect their boundaries.

Holding Time Tips:

- Start with short sessions and gradually increase as trust grows.

- Associate being picked up with positive reinforcement (e.g., treats, affection).

Key Takeaway: Proper handling builds trust and safety. Respect your Ragdoll's body language to foster a stress-free, cuddly bond.

15.2 Encouraging Gentle Handling for Children

Ragdolls' gentle nature makes them ideal companions for families with children, but proper education is vital to ensure a harmonious relationship.

Teach Children the Basics:

- Approach slowly and calmly.

- Always use two hands when petting or holding.

- Avoid loud noises or fast movements that may startle the cat.

- Never pull on the tail, ears, or whiskers.

Age-Appropriate Guidance:

- **Toddlers (Under 4):** Supervised petting only; avoid carrying.

- **Ages 5–8:** Teach respectful petting and the proper way to hold with adult help.

- **Ages 9+:** With training, older children can safely pick up and help care for the cat.

Interactive Activities:

- Encourage reading or quiet play near the cat.

- Use wand toys for play to maintain a safe distance.

Supervision Is Key:

- Always supervise young children around any pet.

- Reinforce positive behavior with praise and gentle reminders.

Key Takeaway: Teaching children empathy and gentleness ensures mutual trust and a lifelong friendship between them and their Ragdoll.

15.3 Reducing Fear During Vet Visits

Vet visits can be stressful, even for easygoing Ragdolls. With preparation, you can minimize anxiety and make the experience more positive.

Familiarization at Home:

- Regularly touch and handle your cat's paws, ears, and mouth to reduce exam stress.

- Use a travel carrier at home for naps or treats so it becomes a familiar space.

The Right Carrier:

- Choose a sturdy, top-loading carrier with soft padding.

- Line it with a familiar blanket or shirt with your scent.

- Spray with cat pheromones (like synthetic feline facial pheromones) before travel.

Travel Preparation:

- Avoid feeding your cat immediately before the visit.

- Play soothing music during the ride.

- Stay calm—your Ragdoll will pick up on your energy.

At the Vet:

- Keep the carrier covered with a towel to block visual stimuli.

- Request feline-friendly exam rooms if available.

- Speak gently and pet your cat during the visit if permitted.

Aftercare Tips:

- Give your cat a quiet space to decompress at home.

- Offer favorite treats or extra affection.

Key Takeaway: Positive associations, calm handling, and the right tools help reduce anxiety for vet visits and build long-term resilience.

15.4 Stress-Free Travel Tips

Whether it's a road trip or a move to a new home, Ragdolls can travel safely if prepared properly.

Pre-Trip Preparation:

- Use a comfortable, well-ventilated carrier

- Acclimate your cat by taking short car rides in advance

- Pack essentials: food, water, litter, toys, and a familiar blanket

During Travel:

- Keep the carrier secure with a seatbelt

- Avoid loud music or erratic driving

- Never open the car doors unless your cat is secure in the carrier

Breaks and Stops:

- For long journeys, stop every 3–4 hours to offer water and check on your cat

- Never leave your Ragdoll alone in a hot or cold vehicle

Hotel or Temporary Stay Tips:

- Set up a quiet, confined area first (e.g., bathroom with litter box and food)

- Allow your cat to explore gradually

- Maintain routines to reduce anxiety

Air Travel Considerations:

- Choose airlines with pet policies suited for cabin travel

- Confirm carrier size requirements and booking logistics well in advance

Key Takeaway: With gradual preparation and thoughtful packing, you can travel with your Ragdoll cat while minimizing stress and keeping them safe.

CHAPTER 16

Safety and Household Hazards

16.1 Common Indoor Dangers for Cats

Your home may seem safe at a glance, but many everyday objects and spaces can pose risks to your Ragdoll cat. Recognizing these potential hazards is essential to creating a secure environment.

Cords and Wires:

- Chewing on electrical cords can cause shock or burns.

- Secure loose cords with protective tubing or hide them behind furniture.

Small Objects:

- Hair ties, rubber bands, coins, and jewelry can be swallowed and cause internal blockages.

- Store small items in sealed containers or drawers.

Appliances and Furniture:

- Cats may hide in washers, dryers, or behind refrigerators.

- Always check before use and block access when not in use.

Windows and Balconies:

- Screens are not always secure; cats can push them out.

- Use cat-proof screens or keep windows closed unless supervised.

Hot Surfaces:

- Stovetops, radiators, and candles can cause burns.

- Cover stovetops when not in use and keep candles out of reach.

Bathrooms:

- Open toilet lids may present drowning risks, especially for curious kittens.

- Keep cleaning products locked away, and avoid open containers.

Key Takeaway: Your home should be cat-proofed with the same care you'd offer a toddler—think ahead, observe, and adapt to their curious nature.

16.2 Toxic Plants and Foods

Many plants and foods that are harmless to humans can be extremely dangerous to cats. Even small amounts can cause severe reactions.

Toxic Plants (Common Examples):

- Lilies (especially dangerous, even pollen or water from a vase is toxic)

- Pothos

- Philodendron

- Aloe Vera

- Sago Palm

- Dieffenbachia

- Tulips, Daffodils, Hyacinths (bulbs especially)

Safe Alternatives:

- Spider plant

- Catnip

- Wheatgrass

- Areca Palm

Harmful Foods for Cats:

- Chocolate (theobromine is toxic to cats)

- Onions, garlic, chives

- Grapes and raisins (can cause kidney failure)

- Alcohol and caffeine

- Raw dough (yeast can expand in the stomach)

- Xylitol (found in sugar-free gum and some peanut butters)

Signs of Poisoning:

- Drooling, vomiting, or diarrhea

- Lethargy or seizures

- Difficulty breathing

- Sudden collapse

What to Do:

- Contact your vet or a poison control hotline immediately

- Bring a sample of the ingested item if possible

Key Takeaway: Prevention is the best approach. Know what's toxic, keep harmful items out of reach, and familiarize yourself with symptoms of poisoning.

16.3 Preventing Accidents in the Home

Creating a safe indoor environment means proactively addressing areas where accidents could happen.

Kitchen Safety:

- Don't leave food unattended on countertops.

- Store sharp tools securely.

- Keep trash bins covered and inaccessible.

Bathroom Tips:

- Secure medications in cabinets.

- Keep razors, floss, and other small items stored away.

- Close toilet lids to avoid risk of drowning or drinking unsafe water.

Bedroom Awareness:

- Be cautious of small items like earrings or sewing supplies.

- Don't leave plastic bags or strings where cats can play unsupervised.

Living Room Dangers:

- Monitor access to lit candles or fireplaces.

- Avoid hanging cords from blinds or curtains.

- Anchor heavy furniture if your cat likes to climb.

Laundry Room and Garage:

- Lock up detergents, antifreeze, and chemicals.

- Ensure no small crawl spaces are accessible.

Tips for Daily Safety:

- Do a quick visual check before leaving home or going to bed.

- Use pet-safe cleaning supplies.

Key Takeaway: A safety-first mindset prevents injury and builds a secure, happy home for your Ragdoll.

16.4 Fire, Emergency, and Evacuation Preparedness

Emergencies can happen without warning. Preparing ahead of time ensures both you and your Ragdoll cat stay safe.

Fire Safety:

- Place window decals on doors or windows indicating you have pets inside.

- Never leave candles or open flames unattended.

- Have smoke detectors in every room, and check them regularly.

Earthquake or Natural Disaster Readiness:

- Secure heavy objects to prevent toppling.

- Keep a flashlight, batteries, and emergency food supply in an accessible location.

Emergency Pet Go-Bag Should Include:

- Copies of medical records

- Extra food and bottled water

- Portable litter box and litter

- Blanket or familiar item

- Carrier and leash/harness

- First-aid supplies (bandages, tweezers, saline wash)

Evacuation Plan:

- Know your nearest pet-friendly emergency shelters or hotels

- Have multiple escape routes mapped

- Keep your cat's carrier accessible at all times

Practice Makes Preparedness:

- Regularly rehearse what you'd do during an evacuation

- Get your cat used to their carrier with treats and short trips

Key Takeaway: Emergencies are unpredictable, but being prepared helps protect your cat's life and gives you peace of mind.

CHAPTER 17

Breeding Basics (Optional Knowledge)

17.1 Understanding Ragdoll Genetics

Ragdoll cats are known for their striking blue eyes, soft semi-long fur, and affectionate demeanor. Behind these features lie complex genetic principles that responsible breeders must understand.

Color and Pattern Inheritance:

- Ragdolls exhibit a range of colors (seal, blue, chocolate, lilac) and patterns (colorpoint, mitted, bicolor).

- Coat patterns are governed by combinations of dominant and recessive genes.

- The blue eye color is closely tied to the pointed gene (temperature-sensitive albinism).

Genetic Health Considerations:

- Hypertrophic Cardiomyopathy (HCM) is a known hereditary condition in Ragdolls; DNA testing is critical before breeding.

- Polycystic Kidney Disease (PKD) can be inherited; genetic screening is advisable.

- Breeding two cats with similar recessive genes can increase the risk of congenital issues.

Pedigree Knowledge:

- Maintaining clear lineage records helps avoid inbreeding and supports healthy genetic diversity.

- Kittens produced by purebred parents should be registered with a recognized cat association.

Key Takeaway: Genetic understanding allows breeders to make informed choices that protect the breed's future health and appearance.

17.2 Responsible Breeding Ethics

Breeding should never be taken lightly. It requires significant time, knowledge, and resources to ensure the welfare of both the parent cats and the kittens.

Essential Ethical Practices:

- Only breed cats that are healthy, temperamentally sound, and cleared by a veterinarian.

- Ensure both the queen and stud are over one year old and fully mature.

- Do not breed a female more than 3 times in 2 years to prevent health exhaustion.

Living Conditions:

- Provide clean, spacious, and calm environments for queens during pregnancy and nursing.

- Socialize breeding cats with humans and other animals to reduce stress.

Avoid Profit-Driven Breeding:

- Ethical breeders focus on quality, not quantity.

- The goal should always be to preserve and improve the breed.

Spaying/Neutering Contracts:

- Many responsible breeders include these clauses in sales contracts to prevent uncontrolled reproduction.

Key Takeaway: Responsible breeding respects the health, welfare, and dignity of every cat involved. Ethical breeding elevates the Ragdoll breed as a whole.

17.3 Pregnancy and Kittening Overview

If breeding is pursued, understanding the pregnancy and kittening stages is crucial for a smooth, safe process.

Signs of Pregnancy:

- Reduced appetite, increased affection, and pinking of the nipples within 2–3 weeks.

- A typical Ragdoll pregnancy lasts 63–68 days.

Pregnancy Care Tips:

- Feed a high-quality, protein-rich diet.

- Minimize stress and provide a quiet space.

- Schedule vet checkups to monitor health and fetal development.

Kittening (Birthing) Preparation:

- Prepare a nesting box in a warm, quiet room.

- Line with soft blankets or towels and keep other animals away.

- Monitor for signs of labor: restlessness, nesting, vocalization.

Birthing Process:

- Most Ragdoll queens deliver kittens without intervention.

- Kittens are usually born 30–60 minutes apart.

- Allow the mother to clean and nurse each kitten; assist only if necessary.

When to Call the Vet:

- Labor lasts more than 24 hours

- Kitten is stuck in the birth canal

- Excessive bleeding or distress

Key Takeaway: Close observation, preparation, and veterinary support are vital during pregnancy and birth.

17.4 Raising Healthy Litters

The first few weeks of a kitten's life are critical for growth, bonding, and immune development.

Neonatal Stage (0–2 Weeks):

- Kittens are blind, deaf, and completely dependent on the queen.

- Monitor weight daily and ensure each kitten is nursing.

Socialization Stage (2–8 Weeks):

- Begin gentle handling around week 3 to support social development.

- Introduce mild household sounds and simple toys.

Weaning (4–8 Weeks):

- Gradually introduce soft food alongside nursing.

- Provide a shallow litter box and safe exploration area.

Health Maintenance:

- Deworming typically begins around week 4.

- First vaccinations between weeks 6–8.

- Monitor for signs of illness: diarrhea, low weight gain, or lethargy.

Rehoming Readiness (12–14 Weeks):

- Kittens should be fully weaned, socialized, and vet-checked before adoption.

- Share a care guide with new owners to ensure continued health and happiness.

Key Takeaway: Proper nutrition, hygiene, and early socialization set the foundation for physically and emotionally healthy kittens.

CHAPTER 18

Veterinary Care and Common Illnesses

18.1 Genetic Predispositions in Ragdolls

While Ragdoll cats are generally known for their good health and robust constitution, like all breeds, they are genetically predisposed to certain medical conditions. Understanding these risks is essential for prevention and early intervention.

Hypertrophic Cardiomyopathy (HCM):

- One of the most common hereditary conditions in Ragdolls, HCM causes the heart walls to thicken, which can lead to heart failure.

- A DNA test is available for breeders to screen for the HCM mutation.

- Regular cardiac screenings are essential, especially if symptoms like rapid breathing, lethargy, or collapse occur.

Polycystic Kidney Disease (PKD):

- Less common in Ragdolls but still a potential concern, PKD results in fluid-filled cysts forming in the kidneys.

- Symptoms may include increased thirst, urination, or weight loss.

- Ultrasound screening can detect the condition early.

Bladder Stones and Urinary Issues:

- Some Ragdolls are prone to urinary tract infections (UTIs) and bladder crystals.

- Signs include straining to urinate, bloody urine, or urinating outside the litter box.

- Ensure your cat drinks plenty of water and receives regular veterinary checkups.

Hereditary Musculoskeletal Issues:

- Joint issues like hip dysplasia, though rare in cats, may occasionally appear in Ragdolls.

- Monitor your cat's movement and activity levels.

Key Takeaway: Genetics play a role in your cat's health. Early testing, regular monitoring, and choosing a responsible breeder help reduce long-term risks.

18.2 Common Illnesses and Treatments

Beyond hereditary issues, Ragdolls may face typical feline illnesses throughout their lives. Recognizing symptoms and acting promptly ensures better outcomes.

Upper Respiratory Infections (URIs):

- Caused by viruses like herpesvirus or calicivirus.

- Symptoms: sneezing, nasal discharge, watery eyes, and lethargy.

- Treatment involves rest, hydration, and sometimes antiviral medications.

Dental Disease:

- Gingivitis and tooth resorption are common in older Ragdolls.

- Symptoms: bad breath, difficulty eating, drooling.

- Treatment includes professional cleanings and, in some cases, tooth extractions.

Parasites (Internal and External):

- Common parasites include fleas, ear mites, and worms.

- Signs: scratching, hair loss, visible worms in stool.

- Routine deworming and vet-recommended parasite preventives are essential.

Diarrhea and Digestive Upset:

- Can result from dietary changes, stress, or infection.

- Persistent symptoms warrant vet evaluation to rule out chronic issues like IBD (Inflammatory Bowel Disease).

Eye and Ear Infections:

- Watch for redness, discharge, or head shaking.

- Clean gently with vet-approved solutions and seek professional care if symptoms persist.

Key Takeaway: Knowing common health issues allows you to respond swiftly. Early detection and professional care are critical to successful treatment.

18.3 Preventing Obesity and Joint Issues

Ragdolls are known for their laid-back, indoor lifestyle, which makes them more prone to weight gain and joint problems if not carefully managed.

Risks of Obesity:

- Obesity can shorten a cat's lifespan and increase the risk of diabetes, arthritis, and liver issues.

- Causes include overfeeding, lack of exercise, and poor diet.

Maintaining a Healthy Weight:

- Feed portion-controlled meals tailored to your cat's age and activity level.

- Avoid excessive treats or free-feeding.

- Weigh your cat monthly to monitor trends.

Joint Health Tips:

- Encourage daily play for muscle tone and flexibility.

- Provide soft bedding and avoid high jumping surfaces for older cats.

- Monitor for signs of stiffness, limping, or reluctance to climb.

When to See a Vet:

- If your cat gains or loses weight rapidly

- Shows signs of pain when walking or being touched

Key Takeaway: Preventive care and daily activity are the best defenses against obesity and joint decline. Small habits have long-term benefits.

18.4 Supporting Long-Term Health

Supporting your Ragdoll's lifelong wellness requires a holistic approach: diet, exercise, emotional health, and regular vet care.

Annual Checkups:

- Schedule wellness exams at least once a year (biannual for senior cats).

- Early diagnosis often makes treatment easier and more successful.

Vaccination Schedule:

- Maintain core vaccines: feline distemper, herpesvirus, calicivirus, and rabies.

- Consult your vet for non-core vaccines based on lifestyle and risk factors.

Routine Screenings:

- Blood work for senior cats to check organ function

- Dental exams and cleanings

- Heart health evaluations, especially for HCM-prone cats

Emotional and Mental Health:

- Provide regular social interaction, play, and mental enrichment.

- Reduce stress through consistent routines and calm environments.

Healthy Environment:

- Clean litter boxes daily

- Provide clean water and a balanced diet

- Keep the home free of hazards

End-of-Life Planning:

- Discuss palliative care options when necessary

- Ensure your cat's final years are filled with comfort, dignity, and love

Key Takeaway: Long-term health is built on consistency—regular checkups, good nutrition, emotional support, and proactive care make all the difference.

CHAPTER 20

Lifespan and End-of-Life Care

20.1 Average Lifespan of Ragdolls

Ragdoll cats are known for their longevity and resilience. With proper care, they can enjoy long, healthy lives alongside their human companions.

Typical Lifespan Range:

- The average Ragdoll lives between 12 and 17 years.

- Some may reach 18–20 years with optimal care, especially when kept indoors and regularly monitored by a veterinarian.

Factors Influencing Longevity:

- **Genetics:** Cats from healthy, well-bred lines are more likely to live longer.

- **Nutrition:** Balanced, age-appropriate diets support immune health and prevent obesity.

- **Preventive Care:** Routine veterinary visits, vaccinations, and early illness detection contribute significantly.

- **Lifestyle:** Indoor-only Ragdolls typically live longer due to reduced exposure to disease, injury, and environmental hazards.

- **Spaying/Neutering:** This can help avoid certain cancers and unwanted behaviors that may reduce lifespan.

- **Dental Care:** Oral health is vital; infections from neglected teeth can affect organs.

- **Hydration:** Ensuring constant access to fresh water prevents kidney and urinary issues.

Environmental Considerations:

- Air quality, access to enrichment, and a stable household environment reduce stress.

- Avoid smoking indoors or using harsh cleaning chemicals near your cat.

- Reduce loud noises and maintain a peaceful home setting.

Mental and Emotional Wellness:

- Mental stimulation through play and interaction helps prevent cognitive decline.

- Loneliness can affect lifespan; companionship matters.

- A bonded human-cat relationship supports emotional health.

Seasonal Care Adjustments:

- Senior cats may be more sensitive to temperature changes. Provide cozy warmth in winter and adequate cooling in summer.

- Monitor for arthritis flares in cold months.

Key Takeaway: While you can't control genetics, excellent care and a safe, enriched environment greatly influence your Ragdoll's lifespan. Loving companionship, attentive healthcare, and stable routines are pillars of longevity.

20.2 Comforting Senior Cats

As your Ragdoll ages, its needs will shift—requiring gentle adaptation to ensure a comfortable and peaceful senior life.

Recognizing Aging Signs:

- Reduced mobility or activity

- Changes in appetite or weight

- Dental issues or bad breath

- Sleeping more or altered behavior

- Decreased grooming or matted fur

- Difficulty jumping or climbing

- Increased vocalization (sign of confusion or discomfort)

- Accidents outside the litter box

Adapting Their Environment:

- Provide orthopedic or low-sided beds for easy access.

- Place ramps or steps to help reach favorite spots.

- Use non-slip mats on slick floors.

- Keep litter boxes easily accessible and with low sides.

- Place food, water, and litter within short walking distance.

Dietary Adjustments:

- Senior-specific nutrition helps with joint support and organ health.

- Increased hydration support (wet foods or water fountains) aids kidney function.

- Smaller, more frequent meals may be easier to digest.

- Supplements such as omega fatty acids (if approved by your vet) can help joints.

Routine Monitoring:

- Biannual vet visits to catch early signs of disease

- Monitor vision, hearing, and cognitive health

- Blood tests to assess kidney, liver, and thyroid function

- Weigh-ins to detect subtle weight loss or gain

- Dental exams and cleanings to prevent infections

Emotional Comfort:

- Keep routines predictable

- Offer extra affection and quiet time

- Reduce stress by limiting household changes

- Use pheromone diffusers to create a calming environment

Mental Enrichment:

- Provide soft puzzle toys or gentle interactive games

- Allow time near sunny windows or bird-watching stations

- Gentle brushing sessions also help with bonding

Support for Incontinence:

- Add washable pee pads around the litter area

- Use extra litter boxes with easy access and non-clumping litter

- Clean soiled fur with warm damp cloths or cat-safe wipes

Hygiene Support:

- Assist with grooming if flexibility declines

- Check eyes and ears for signs of infection

- Trim nails more frequently as seniors are less active

Key Takeaway: Small lifestyle changes can dramatically improve your senior cat's quality of life—both physically and emotionally. Observation, love, and gentle care go a long way.

20.3 Palliative and End-of-Life Support

When your Ragdoll enters the final stages of life, compassionate, proactive care is essential to reduce suffering and preserve dignity.

What is Palliative Care?

- Focuses on comfort and quality of life when a cure is no longer possible

- May include pain management, hydration support, and special diets

- Encourages continued bonding time and emotional support

- Includes mobility support, incontinence management, and temperature control

Signs It May Be Time for End-of-Life Planning:

- Chronic pain that's no longer manageable

- Loss of mobility, interest in food, or interaction

- Frequent illness or hospitalization

- Confusion or distress that cannot be soothed

- Soiling themselves or inability to rest comfortably

- Labored breathing or vocalizing in pain

Working with a Veterinarian:

- Your vet will guide you through hospice care options

- Discuss humane euthanasia when suffering outweighs comfort

- Pain medications and anti-anxiety treatments may improve quality of life temporarily

- Your vet can teach you how to assess quality of life using scoring charts

Home Support:

- Create a soft, warm resting area in a quiet, familiar location

- Use padded bedding to prevent sores from inactivity

- Offer hand-feeding or assist with grooming when needed

- Dim lighting and soft music may help ease anxiety

- Keep the cat's litter box clean and close by

- Keep food and water close, and gently hand-feed if necessary

Hydration and Nourishment:

- Use shallow dishes to encourage drinking

- Offer broths or flavored water if needed

- Warm up food slightly to make it more appealing

- Use a syringe (as guided by a vet) for fluids or food if needed

When Euthanasia is the Kindest Choice:

- It's a deeply personal decision best made with veterinary input

- Prioritize your cat's comfort, dignity, and emotional peace

- It's okay to grieve before and after the decision

- Some vets offer at-home euthanasia for a more peaceful goodbye

Post-Euthanasia Options:

- Burial (home or pet cemetery)

- Cremation (private or group)

- Keepsake urns, paw prints, or memorial jewelry

- Memorialization through scrapbooks or digital tributes

Support During This Time:

- Keep close friends or family informed

- Let others help with tasks so you can be present with your pet

- Don't hesitate to cry or express emotion

- Consider writing a letter of gratitude to your pet

Key Takeaway: Providing gentle, attentive care during your Ragdoll's final chapter is the greatest gift you can offer. Celebrate their life while ensuring comfort and dignity.

20.4 Coping with Loss and Grief

Losing a pet—especially one as affectionate and present as a Ragdoll—can be profoundly painful. Understanding grief and allowing yourself to mourn is a healthy part of the healing process.

Grieving Process:

- Emotions can include sadness, guilt, anger, and loneliness

- Grief has no set timeline; everyone heals differently

- Don't feel pressured to "move on" quickly

- Pets are family—mourning them is valid and necessary

- You may experience grief in waves—this is normal

Healthy Coping Strategies:

- Create a tribute or photo album

- Write a letter to your pet or journal your memories

- Speak openly with understanding friends or support groups

- Light a candle, plant a tree, or hold a small ceremony

- Visit the pet's favorite resting place

- Look through happy photos and remember the joy they brought

Supporting Children and Family:

- Be honest and gentle when explaining pet loss

- Allow children to express their feelings and ask questions

- Let them create a goodbye letter or draw a picture for closure

- Include them in memorial rituals to give them a sense of closure

Pet Loss Support Resources:

- Grief hotlines or support groups

- Pet loss counselors or online forums

- Memorial services or ceremonies

- Books and resources on pet bereavement

- Ask your vet if they recommend any local support networks

Workplace and Social Considerations:

- Consider taking a day off if needed

- Communicate with your employer if you're grieving deeply

- Decline social invitations temporarily if emotionally overwhelmed

- Ask for help if daily tasks feel overwhelming

Deciding on a New Pet:

- It's okay to wait—or to never adopt again

- When ready, honor your lost pet's legacy by giving love to another in need

- Some families find healing in fostering before deciding on adoption

- Your bond with your lost pet is not replaced—it's honored by future love

Keeping Their Memory Alive:

- Display a framed photo or paw print keepsake

- Share stories of your pet with others who loved them

- Donate in their honor to an animal rescue

- Write about your experiences to help others cope with loss

- Create a memorial corner with their favorite toy or blanket

Key Takeaway: Mourning a beloved Ragdoll is valid and necessary. Grief reflects the deep bond shared—and healing takes time and support. You are not alone in your sorrow, and it's okay to honor that love in your own way.

CHAPTER 21

Behavioral Challenges

21.1 Separation Anxiety Solutions

Ragdolls are affectionate, people-oriented cats that often form strong bonds with their human companions. This attachment, while endearing, can sometimes lead to separation anxiety when their favorite person is away.

Understanding Separation Anxiety in Ragdolls:

- Exhibiting stress behaviors when left alone

- Excessive vocalization or meowing

- Destructive behavior (scratching furniture, knocking items down)

- Litter box issues (urinating outside the box)

Prevention and Management Tips:

- Gradually increase alone-time periods to build tolerance

- Create a consistent leaving and returning routine

- Use puzzle feeders and treat-dispensing toys to provide stimulation while you're away

- Play calming music or leave a piece of clothing with your scent

Enriching the Environment:

- Provide window perches for bird-watching

- Place interactive toys around the home

- Use motion-activated toys or videos made for cats

Professional Support:

- Consult a feline behaviorist for severe cases

- Consider pheromone diffusers to reduce anxiety

Key Takeaway: Addressing separation anxiety helps build a confident, independent cat while maintaining their emotional security.

21.2 Aggression and Biting

While Ragdolls are generally gentle and laid-back, occasional aggression or biting can occur due to stress, fear, or health issues.

Common Causes of Aggression:

- Fear or overstimulation

- Sudden environmental changes

- Pain or illness

- Poor early socialization

- Redirected aggression (e.g., seeing another cat outside)

Types of Aggression:

- **Play Aggression:** Common in kittens; includes biting during play

- **Territorial Aggression:** Often directed at new pets or people

- **Fear-Based Aggression:** Triggered by unfamiliar stimuli

Managing and Preventing Aggression:

- Avoid rough play that encourages biting

- Watch for warning signs: tail flicking, ears back, dilated pupils

- Provide structured playtime to burn off energy

- Ensure your cat has safe spaces to retreat to

When to Seek Help:

- If biting draws blood or occurs regularly

- If aggression disrupts household harmony

- Veterinarians can rule out medical causes

Key Takeaway: Identifying the root of aggression allows you to address it with targeted strategies, improving trust and safety.

21.3 Over-Grooming and Stress Behaviors

Over-grooming is a common stress behavior in cats and may signal physical discomfort or emotional imbalance.

Recognizing Over-Grooming:

- Bald patches or thinning fur

- Licking a specific area repeatedly

- Red or irritated skin

Possible Triggers:

- Stress or anxiety (e.g., new pets, moving homes)

- Allergies (environmental or food-related)

- Parasites (fleas, mites)

- Pain (especially in joints or stomach)

Addressing the Issue:

- Consult your vet to rule out medical conditions first

- Use calming pheromones or sprays

- Enrich the environment with more stimulation and hiding spaces

- Avoid punishing the behavior—it increases stress

Encouraging Healthy Grooming:

- Regular brushing helps distribute oils and reduce stress

- Engage in daily play to redirect compulsive energy

- Keep a predictable routine to lower anxiety levels

Key Takeaway: Over-grooming is often a symptom of something deeper. Identifying and treating the root cause ensures long-term well-being.

21.4 Litter Box Avoidance Fixes

Inappropriate elimination is one of the most common behavioral challenges faced by cat owners—but it's also one of the most solvable.

Common Causes:

- Dirty or infrequently cleaned litter boxes

- Box location too busy or loud

- Change in litter type or scent

- Medical issues (e.g., urinary tract infection)

- Stress or territorial behavior

Solutions and Preventative Measures:

- Scoop litter at least once daily; clean the entire box weekly

- Have one litter box per cat, plus one extra (especially in multi-cat homes)

- Place boxes in quiet, easily accessible areas

- Stick to unscented litter that mimics natural substrates

Medical Considerations:

- A sudden change in litter habits may indicate illness

- Schedule a vet visits to rule out infections, bladder stones, or arthritis

Behavioral Support:

- Clean soiled areas thoroughly with enzymatic cleaner

- Never punish or scold—redirect with praise when using the box correctly

- Try different litter box styles: open, covered, or low-sided

Key Takeaway: Litter box problems often reflect a cat's attempt to communicate. Listening and adjusting ensures harmony in the home.

CHAPTER 22

Advanced Training and Enrichment

22.1 Teaching Tricks and Commands

While cats are often seen as independent and less trainable than dogs, Ragdolls are uniquely receptive to learning tricks and commands thanks to their affectionate and intelligent nature.

Benefits of Trick Training:

- Enhances mental stimulation

- Strengthens the human-cat bond

- Builds confidence and reduces boredom

Start with Basic Commands:

- **"Sit":** Use a treat to guide your cat's head back and reward when they sit

- **"High Five":** Tap their paw and reward the behavior

- **"Come"**: Use their name and reward when they respond

Training Tips:

- Use positive reinforcement (treats, praise, petting)

- Keep sessions short (5–10 minutes) and consistent

- Train when your cat is alert but not overly energetic or hungry

- Use clicker training to mark correct behaviors

Common Mistakes to Avoid:

- Never use punishment—it causes fear and resistance

- Avoid overwhelming your cat with too many commands

Key Takeaway: Even basic tricks can enrich your Ragdoll's life and deepen your bond.

22.2 Harness and Leash Training

While most cats stay indoors, harness and leash training can open up safe, supervised outdoor experiences that stimulate their senses and boost happiness.

Why Train Your Cat to Use a Harness:

- Safe exposure to the outdoors

- Helps reduce boredom and stress

- Great for physical activity and confidence building

Choosing the Right Harness:

- Use a well-fitted, escape-proof harness

- Avoid collars or loose-fitting gear

Training Process:

1. **Familiarize Indoors:** Let your cat sniff and inspect the harness

2. **Practice Wearing:** Put the harness on briefly indoors and offer treats

3. **Add Leash:** Attach the leash indoors and allow your cat to explore

4. **Short Outdoor Walks:** Start in a quiet yard or patio

Tips for Success:

- Be patient and never rush the process

- Always supervise outdoor walks

- Avoid noisy or crowded areas until your cat is fully confident

Key Takeaway: Leash training is achievable and rewarding with time, consistency, and gentle encouragement.

22.3 Mental Enrichment Exercises

Mental stimulation is just as important as physical play for Ragdolls, especially given their curious and affectionate nature.

Enrichment Benefits:

- Prevents boredom-related behaviors

- Keeps cognitive function sharp

- Provides daily structure and engagement

Enrichment Activities:

- **Puzzle feeders:** Encourage problem-solving for food

- **Hide-and-seek games:** Hide toys or treats for them to find

- **Interactive toys:** Toys that move, squeak, or dispense rewards

- **Training sessions:** Learning tricks can double as mental enrichment

- **Scent work:** Let your cat sniff different fabrics, spices, or herbs (safe ones like catnip or silvervine)

DIY Enrichment Ideas:

- Cardboard box mazes

- Tunnels or hideouts made from household items

- Frozen treat toys using safe ingredients

Rotating Toys:

- Change toys weekly to keep them exciting and fresh

Key Takeaway: Mental challenges are essential for well-being and can be easy to integrate into daily routines.

22.4 Problem-Solving for Stubborn Cats

Even the most affectionate Ragdolls can show signs of stubbornness or resistance to new routines and behaviors. Understanding and adapting is key.

Why Stubbornness Happens:

- Change in routine or environment

- Lack of motivation or boredom

- Miscommunication between cat and owner

- Previous negative experience

Strategies for Encouraging Cooperation:

- Break tasks into smaller steps

- Use high-value rewards during training

- Stick to consistent cues and routines

- Avoid forcing interactions or rushing training

Reading the Signs:

- Flattened ears, tail twitching, avoidance = stress or resistance

- Soft purring, tail up, head bumps = openness and comfort

Building Trust Over Time:

- Respect boundaries and use gentle handling

- Use play as a bridge to cooperation

- Reward all progress—even small wins

Key Takeaway: Patience, consistency, and compassion help even the most independent Ragdoll become a cooperative companion.

CHAPTER 23

250+ Bonus Tips, Hacks, and Fast Facts

23.1 Fun Ragdoll Trivia

1. Ragdolls are called "puppy cats" due to their dog-like affection.

2. They often go limp when picked up—hence the name "Ragdoll."

3. All Ragdoll kittens are born white; their coat color develops as they age.

4. The breed was developed in the 1960s in California by Ann Baker.

5. Ragdolls are one of the largest domestic cat breeds.

6. Their eye color is always blue—ranging from light to deep sapphire.

7. Ragdolls are a relatively quiet breed with soft, gentle voices.

8. They prefer human companionship and may follow their owners around.

9. They are often receptive to leash training.

10. Ragdolls are slow to mature and may not reach full size until age 4.

23.2 Quick Health and Grooming Hacks

11. Use a wide-toothed comb followed by a finer brush to prevent tangles.

12. Groom during cuddle time to make it feel rewarding.

13. Use dampened cotton pads for quick eye cleaning.

14. Dental wipes can help keep teeth cleaner between professional cleanings.

15. Brushing once every 2–3 days prevents most mats.

16. Keep nails trimmed monthly to avoid snags on furniture.

17. Place a non-slip mat under water bowls to avoid spills.

18. Use shallow dishes to reduce whisker fatigue.

19. Regularly inspect ears for wax buildup or foul odors.

20. Hydration boosts skin and coat health—offer running water sources.

23.3 Everyday Time-Saving Tricks

21. Keep all grooming supplies in one labeled container.

22. Feed meals at the same time each day to reinforce habits.

23. Use auto-feeders to support routine and convenience.

24. Scoop litter daily to avoid long weekend cleaning sessions.

25. Set reminders for vet visits and grooming days.

26. Use rubber gloves to collect fur from furniture quickly.

27. Freeze small towels to place in beds during summer heat.

28. Schedule play sessions before bedtime to encourage better sleep.

29. Rotate toys weekly to prevent boredom.

30. Create a care calendar with tasks divided weekly/monthly.

23.4 Enrichment and Bonding Boosters

31. Place a bird feeder outside a safe window perch.

32. Use boxes and tunnels for indoor obstacle courses.

33. Offer puzzle feeders for mealtimes.

34. Include cat-safe herbs like catnip or silvervine in play.

35. Mirror play or chase-the-light games boost physical activity.

36. Clicker train for mental stimulation and bonding.

37. Use treat-hiding toys to keep your cat engaged.

38. Practice daily brushing as a bonding activity.

39. Use gentle, calm tones to communicate consistently.

40. Allow quiet time in your lap—even if your Ragdoll isn't cuddly at the moment.

...and over 200 more tips, fast facts, and hacks will be peppered throughout the book and summary recap to help you easily apply them in daily life!

From fun facts to practical hacks, this chapter ensures your Ragdoll journey is not only easier but more joyful and deeply connected.

CHAPTER 24

Free 30-Day Care & Training Plan

This beginner-friendly, 30-day care and training roadmap is designed to guide first-time Ragdoll owners through a smooth, successful introduction to life with their new feline companion. The goal is to establish healthy routines, deepen your bond, and build trust from the very beginning.

Each week is thoughtfully structured to support your Ragdoll's physical, emotional, and behavioral development while reinforcing responsible pet parenting habits.

24.1 Week 1: Settling in and Bonding

Theme: Safety, comfort, and initial trust-building.

Day 1–2: Welcome & Environment Setup

- Set up a small, enclosed room or area with all essentials: food, water, litter box, cozy bedding, and a few safe toys.

- Avoid overwhelming your Ragdoll with the entire house right away.

- Sit quietly nearby—read aloud, talk softly, or simply be present to allow them to acclimate to your scent and voice.

Day 3–4: Slow Introduction to Routine

- Begin feeding at regular times to establish structure.

- Gently introduce interactive play (wand toys or feather chasers) and scratching surfaces.

- Let your cat explore other rooms gradually and always at their own pace.

Day 5–6: Encouraging Touch and Affection

- Start with short petting sessions on safe zones: chin, cheeks, and back.

- Introduce a soft-bristle grooming brush, rewarding calm behavior with treats.

- Keep sessions brief but consistent, always stopping if your cat shows signs of discomfort.

Day 7: Strengthening Emotional Connection

- Set aside 15–30 minutes for intentional bonding: play, brushing, or lap time (if accepted).

- Reinforce eye contact and verbal cues gently.

- Start developing a predictable rhythm: wake-up, feed, play, nap, evening bonding.

24.2 Week 2: Feeding, Grooming, and Play

Theme: Routine-building and physical care.

Day 8–9: Fine-Tuning Nutrition

- Monitor eating habits: amount consumed, preferences, digestion.

- Offer variety using natural food options or healthy meal toppers (no additives or commercial promotions).

- Keep a food journal to track reactions and establish dietary preferences.

Day 10–11: Grooming Familiarity

- Introduce grooming as a bonding experience, not a chore.

- Brush in short intervals, speaking softly throughout.

- Begin gentle ear and eye cleaning using damp cloth or cotton pads.

- Handle paws and extend nails for desensitization—no clipping yet if your cat resists.

Day 12–13: Daily Play Integration

- Use toys that simulate prey movement to activate your Ragdoll's hunting instincts.

- Try 2–3 play bursts of 10 minutes each throughout the day.

- End each session with a reward or calm cuddle to transition from excitement to relaxation.

Day 14: Full Grooming Session & Comfort Check

- Perform a complete grooming session: brushing, ear check, eye cleaning, and paw handling.

- Introduce a small grooming mat or routine area to create positive association.

- Use this time to assess body condition, coat health, and any behavioral discomfort.

24.3 Week 3: Training and Socialization

Theme: Foundations for lifelong learning and confidence.

Day 15–16: Name Recognition & Engagement

- Use your cat's name in positive, low-pressure contexts (e.g., feeding, play).

- Reward every time your cat looks at you or responds.

- Begin associating their name with call-and-respond games using treats.

Day 17–18: Litter Box Mastery

- Monitor consistency of use and cleanliness.

- Remove waste daily and fully change litter every 5–7 days.

- Praise or offer calm reassurance when your cat uses the box appropriately.

Day 19–20: Basic Trick Training

- Teach simple cues: "come," "sit," or "touch."

- Use a clicker or verbal marker (e.g., "yes!") paired with a treat.

- Limit training sessions to 5–10 minutes to avoid fatigue.

Day 21: Controlled Social Exposure

- Allow your Ragdoll to meet a calm friend or family member.

- Offer treats and respect retreating behavior.

- Create a neutral "safe zone" they can return to freely.

24.4 Week 4: Building Long-Term Routines

Theme: Independence, consistency, and lifelong wellness.

Day 22–23: Mental & Sensory Enrichment

- Rotate toys and introduce puzzle feeders to challenge problem-solving skills.

- Create visual stimulation by placing a cat tree near a window or using scent-based exploration games.

- Offer new textures like crinkle mats, fleece, or felt to explore.

Day 24–25: Developing Solo Confidence

- Allow your Ragdoll quiet solo time in another room.

- Resist the urge to follow—trust helps foster independence.

- Leave behind a shirt with your scent to ease separation.

Day 26–27: Weekly Health Routine

- Perform a gentle physical check: teeth, ears, coat, nails, eyes, hydration.

- Begin documenting any changes or concerns in a pet care log.

- Make notes on eating, litter habits, mood, and behavior.

Day 28–29: Harness & Outdoor Readiness (Optional)

- Introduce harness indoors without the leash at first.

- Reward calm behavior with treats or praise.

- Once accepted, practice walking short distances indoors before considering a controlled outdoor experience.

Day 30: Reflection, Review & Celebrate

- Review the last 30 days: What worked? What needs adjustment?

- Reflect on behavioral milestones, trust-building progress, and routine success.

- Celebrate with a new toy, enrichment treat, or extra cuddle session.

- Start building your monthly maintenance schedule: grooming, vet checks, play, training.

Final Note:

This 30-day plan isn't just about routines—it's about building the lifelong foundation of a calm, bonded, and confident cat. As your Ragdoll's personality unfolds, you'll find that the time you invest now leads to an incredibly rewarding and affectionate partnership for years to come.

CHAPTER 25

Conclusion & Key Takeaways

25.1 Lessons Learned from Ragdoll Ownership

Owning a Ragdoll cat is more than just having a pet—it's an immersive, heartwarming experience that teaches patience, mindfulness, and deep empathy. These gentle giants reveal just how profound the bond between humans and cats can become when nurtured with kindness and understanding.

From the beginning of this guide, we've explored the many facets of Ragdoll ownership:

- **Understanding the breed's history and unique temperament** equips you to care for your Ragdoll with respect for its lineage and instincts.

- **Preparing your home** properly from day one ensures a seamless and stress-free adjustment for your new feline family member.

- **Grooming and feeding routines** build predictability and trust.

- **Training and socialization** open the door for communication, engagement, and mutual respect.

- **Mental stimulation and play** enhance your cat's daily happiness and help prevent destructive or anxious behaviors.

- **Health maintenance** promotes longevity and vitality, allowing you to enjoy your cat's companionship for many beautiful years.

Perhaps the most important lesson of all is that **Ragdoll ownership is a journey of connection**. Every brush stroke, cuddle session, or playful exchange builds emotional trust and affection.

25.2 The Joys of Living with Gentle Giants

Living with a Ragdoll cat is like inviting calm, warmth, and loyalty into your home. These cats aren't just beautiful—they're emotionally intuitive, affectionate, and subtly expressive.

Daily joys include:

- Waking up to soft meows and sleepy eyes peering at you from the edge of the bed

- Sharing a quiet moment while they curl up beside you during a book or movie

- Watching them elegantly stretch or roll over for a belly rub, full of trust

- Laughing at their gentle curiosity as they observe everything in their world

- Witnessing their patience with children and their acceptance of other animals

Unlike some more independent or aloof breeds, Ragdolls thrive on human interaction. They are not only companions but also emotional anchors. Whether you live alone or in a bustling family home, a Ragdoll cat adapts with grace and charm.

25.3 Final Care Reminders and Essentials

As you continue your journey with your Ragdoll, keep these essential practices in mind:

- **Consistency is Comfort:** Cats thrive on routines. Regular feeding, grooming, and playtime reduce anxiety and build trust.

- **Observation is Prevention:** Monitor your Ragdoll's behavior, appetite, and litter box habits closely. Early detection is key to addressing health issues.

- **Gentle Grooming:** Their long, silky coat needs care. Brush every few days to prevent tangles and create bonding opportunities.

- **Enrichment is Vital:** Rotate toys, create window-viewing spots, and engage in interactive play to keep their minds sharp.

- **Respect Their Voice:** Ragdolls may not be loud, but they communicate clearly through behavior, body language, and soft vocal cues.

- **Provide Safe Spaces:** Whether it's a cat tree, hammock, or cozy corner, allow them places to retreat and relax undisturbed.

- **Vet Visits Are Non-Negotiable:** Schedule annual exams, vaccinations, and dental checkups to ensure your cat stays healthy.

Caring for a Ragdoll is as much about emotional connection as it is about physical maintenance. Respect, consistency, and love are your most valuable tools.

25.4 Words of Encouragement for New Owners

Starting out with a new pet—especially a Ragdoll—can feel both exciting and intimidating. But take heart: every experienced cat owner began where you are now.

Here are some final thoughts to carry with you:

- **Every cat is unique.** Don't compare your Ragdoll to others; instead, learn who *they* are. Let them reveal their personality in their own time.

- **Mistakes are part of learning.** You might forget a grooming session or miss a litter box cleanup. That's okay. What matters is growth, not perfection.

- **Bonding takes time.** True trust isn't built overnight. A slow blink, a soft purr, or a nudge from their head means more than words can say.

- **You're not alone.** There are Ragdoll communities, forums, and support groups filled with people just like you—loving, learning, and growing with their feline companions.

- **The effort is worth it.** The love and companionship you'll receive in return from your Ragdoll will be one of the most rewarding experiences of your life.

You are now equipped with the knowledge, tools, and inspiration to provide the very best life possible for your Ragdoll. The chapters of this book may end here—but the journey of shared love, play, care, and trust is just beginning.

Thank you for dedicating your time to understanding and embracing the Ragdoll breed. May your home be filled with soft pawsteps, warm purrs, and a lifetime of unconditional love.

Printed in Dunstable, United Kingdom